The Sensible Prepper

The Sensible Prepper

PRACTICAL TIPS FOR EMERGENCY PREPAREDNESS AND BUILDING RESILIENCE

Michelle and Cam Mather

Tamworth, Ontario Canada K0K 3G0

Library and Archives Canada Cataloguing in Publication

Mather, Cam, 1959-, author
The sensible prepper : practical tips for emergency preparedness and building resilience / Cam Mather and Michelle Mather.

Includes bibliographical references.
Issued in print and electronic formats.
ISBN 978-1-927408-05-6 (pbk.).--ISBN 978-1-927408-06-3 (html)

1. Emergency management. 2. Preparedness. 3. Self-reliant living.
I. Mather, Michelle, 1960-, author II. Title.

GF86.M38 2013 613.6'9 C2013-902580-4
C2013-902581-2

Acknowledgements

We are grateful to the many people who have helped us with this book. We are extremely indebted to Heidi Lind and Ellen Horak who took the time to proof the first draft and offer their comments. We'd also like to thank everyone who has bought our books in the past and continues to support us in our many endeavors. We really appreciate it!

We would like to dedicate this book to our daughters who have brought us so much joy in life, and who continue to enrich our lives along with their husbands and their families.

Table of Contents

PART I
Setting the Stage

1 Introduction

"The End is nigh," or so you might think watching the news these days. Superstorm Sandy wreaked havoc on the U.S. East Coast, leaving millions without power, homes, and more than $60 billion in damage. Brutal tornados destroyed whole towns like Joplin, Missouri. Wild fires ravaged huge tracts of land and homes in the summer of 2012. The drought was epic and devastated much of North America. Shipping on the Mississippi River was in jeopardy because of a lack of rain. And the heat! July 2012 was the hottest month ever, topping off 327 consecutive months that were hotter than the 20th Century global average. Then winter 2013/14 brought the polar vortex and record cold. At one point snow was on the ground in 49 of 50 states. Cars were stranded on freeways in Atlanta Georgia because of ice. And while North Americans kept having freezing winters, 2014 was the warmest year on record worldwide when you look at global average temperatures. Yikes!

Something's definitely happening, but many of us just can't put our finger on it. The world price of oil remained close to $100/barrel even while the economy was sluggish which had destroyed demand and should have pushed down the price. But then it did drop to $50/barrel making much of the shale oil and tar sands oil economically unfeasible. Water is becoming as important as oil in many parts of the country and the world at large, especially those plagued by drought, or trying to grow food. Natural gas is cheap but we keep hearing about how the fracking that is helping to increase supply is contaminating the water supply for many rural dwellers and potentially causing earthquakes while leaving ponds of toxic wastewater. And still we add 50 million more mouths to feed each year to the planet's population, which has already passed the 7 billion mark. This can't be good. Sometimes it feels like something has to give. Sometimes it feels like it really is the end of the world as we know it.

What's pretty obvious to most people is that things have changed, and will continue to change quite quickly. The concept of a stable static world is long past and it would appear the pace of change is increasing at warp speed, and not necessarily in a positive direction.

The other thing that many people are sensing is that our governments, the people we elect to represent us and manage the big picture, have lost the ability to do so. Partisan politics is making it impossible to take any bold moves just at a time when we really need some big ideas to deal with things. And if they just had to deal with one of the problems we're talking about, then maybe we'd have some hope they could make some progress. The problem is that the problems are all happening simultaneously. Peak oil, resource depletion, a warming and changing climate, water shortages, drought, wild fires exacerbated by insect infestations like pine beetles, spiking food prices, and on and on.

It might not be so bad if our governments were flush with cash, but they're broke. Sure they can keep printing money but that's just making the problem worse, increasing inflation and handicapping their ability to direct resources where they're needed.

Probably the best example of what we might want to expect from governments in the future comes from the September 2009 edition of "The Atlantic" magazine in an article about Craig Fugate, the new head of the Federal Emergency Management Agency or FEMA. Most of us remember his predecessor, "Brownie," being complimented by President Bush for the great job he was doing, while thousands of residents of New Orleans were still in desperate conditions following Hurricane Katrina. Fugate says, "Who is going to be the fastest responder when your house falls on your head? Your neighbor."

I think what the government is basically telling us here is that we're on our own. They will show up eventually, but the days of assuming they will arrive quickly, feed you, keep you warm and rebuild your house are over. It took the U.S. Congress 10 days to allocate money for New Orleans after Hurricane Katrina. In the squabbling over the 2012 "Fiscal Cliff" crisis, Congress still hadn't set aside money for the victims of Superstorm Sandy 70 days after the devastation.

One of the candidates for President of the United States in 2012, Mitt Romney, could even be considered a "prepper." He is a Mormon and part of their faith dictates that they have a one-year supply of food in their home. I believe there is profound wisdom in this practice. So it's time you had a "Plan B." And a "Plan C."

There are lots of "prepper" books, books for people who want to prepare for an uncertain future. So what makes this book different?

First off, this book was written by someone who lives pretty independently all of the time, not just during times of crisis. I have lived off the electricity grid for 15 years, making my own electricity, heating my home with wood from my property and I grow a lot of my own food. I don't grow it all and frankly I don't want to because I like oranges and coffee, neither of which would survive in my northern growing zone... yet.

I think this has been one of the biggest challenges for people reading most prepper books. The books are hardcore. They are written by people who almost seem to want chaos to descend on humanity. I have two daughters and I do not want that for them. I'm also not convinced that things will get that bad, that quickly. They could, but what if they don't? It seems kind of an overreaction to barricade yourself in some bunker when there are still so many wonderful things to experience in the world today.

There have always been naysayers calling for the end of the world as we know it. Many people have experienced these sorts of cataclysmic events. Tens of millions of Europeans were dislocated during World War II, a time when a very advanced and civilized part of the world was thrown into convulsions of disruption and disorder. Millions of people were dislocated after the tsunami devastated huge areas around the Indian Ocean in the winter of 2004. More than 220,000 people died but others saw their whole village and world for that matter, wiped out. We watched the tsunami hit the coast of Japan in 2011 destroying whole towns. So major events do happen that change people's lives all the time. The challenge is how you respond.

In recent years the doomers have turned up the volume. It started with the "Y2K" computer glitch that threatened to throw civilization into turmoil on New Year's Eve 2000. There were many problems but nothing as bad as had been predicted.

Since the mid 2000s many doomers have been suggesting that peak oil would send advanced civilization back into the middle ages and we would all be shivering in the dark. So far the lights have stayed on. I do believe we hit peak oil in about 2005 but I also believe humans are very resourceful. For years I have followed many blogs by doomers. They follow a common thread. "I would expect by Labor Day there will be rioting in the streets." Shortly after an uneventful Labor Day we read that "the wheels are most likely going to come off before Christmas" and yet there we all are watching college football games on New Year's Day.

I'm not suggesting that there aren't jarring social changes going on behind the scenes. There are. Every day. People lose their jobs. People lose their pensions and security. People discover they can't afford health saving procedures. Some people definitely are experiencing great challenges. But so far society as a whole has been able to keep the veil of civilization in place and keep moving forward.

I believe though that the challenges we face are eventually going to make this forward momentum stall. I think we will hit a point where disruption will be the norm and the good ole days will start to feel like a fond daydream.

The thing is that many of us have the resources right now to start creating a "Plan B," but we are easily distracted. There's always something more fun to do with our time. There's always something cooler to buy to make us happy. This prepping stuff can be a downer.

I can't help you there too much, but I will say this. Everything I'm recommending will make you more independent. It will mean that next time a tornado or ice storm or hurricane takes out the power in your community, your lights should stay on. It means that next time there is a spike in the price of home heating oil, you won't watch your bank account plummet. And it means the next time that someone in your household is laid off, there will be many months of meals in the pantry to help you over that bump in the road.

And you know what? You'll really start getting into this whole independence thing. It will become a self-reinforcing process, because once you realize how gratifying it is to increase your resilience to shocks the system may experience, you'll find yourself taking this to the next level instinctively. It becomes a positive feedback loop which feeds on itself.

To do this we'll take a look at the two perspectives of the challenge. The first is short term. How do you prepare for that next power outage? What should you take with you if you're forced to flee an approaching hurricane? These are simple, easy tasks and most of us know we should have a family disaster plan in place. But we don't. Some sort of inertia keeps us from getting cracking. Let me help you take that first step.

Then we'll look at the long term and the big picture. Over time energy and food will be getting more expensive, probably much more. And storms will become more intense. So where should we live? How should we heat and power our home? How do we stay safe?

And most of all, how do we prepare our families so that we can be the first responders for our neighbors? Ultimately the greatest safety comes from community. The more prepared you are personally, the better able you'll be to help others. This is good for the community, and it's good for your soul. It's one of the things that makes us human and can give us the greatest joy.

When you're done reading this book you'll have a plan that will help you prepare your family for a disaster that could hit just about any time now. You'll have the knowledge to start making choices about some of those big picture decisions we come up against. What should we replace our furnace with when our current one packs it in? Do we really need a generator? Where should we be looking to live in the future? Do we really need a root cellar?

Humans have always lived with uncertainty. In fact, it's only in the last few generations that we've had the luxury of reducing much of that uncertainty and farming out all the essential services that we need. Someone else grows our food. Someone else provides the fuel to heat our home. Someone else keeps the lights on. Someone else gives us our water. A few generations ago our grandparents did all of this stuff themselves. If there was an ice storm they might have spilled some milk walking back from the barn, but the farmhouse would be warm, they would have water from their well, they would have vegetables in the root cellar, they would have eggs from their chickens and if they needed to, they'd be able to head to the barn and bring back meat for dinner.

Today an ice storm, power outage or any disruption in any of the services we need to survive, is a disaster for most people. It's a catastrophe. Images of people standing outside of homes on the New Jersey coast after Hurricane Sandy, on a cold November day, with no heat, no water, no electricity, no food, were heart-wrenching to watch. But that doesn't have to be you. You have a choice. You have the resources to create a plan to ensure that life will continue on, maybe not as great as it was, but it certainly will be livable.

I suggest you make your grandparents proud. Come up with a plan to make sure that the bumps in the road that the future has in store for you are just that, bumps to get over, not cliffs that you fall off. You have the resources. You most likely have the income and skills. You just need to reach down and find that independent self. It's there, ready to spring into action. Now let's just get to the tools you'll need!

2 The Accidental Preppers

My wife Michelle and I live on 150 acres in the woods in Eastern Ontario in Canada. During our first seven years here our daughters, who are now 26 and 28, lived with us. They've both since moved out and are now living in the city. We are surrounded by thousands of acres of bush that are partly owned by groups of men who use it for two weeks of the year for deer hunting, and the rest is an undeveloped provincial park. Our nearest neighbors are 2 ½ miles (4 km) to the east and 4 miles (6 km) to the west. This happens to be where the electricity poles end as well, so we are completely off the electricity grid and generate all our own electricity with solar and wind and an ever-decreasing amount of fossil fuel in the form of gasoline that we burn in our generator.

We moved here from a busy suburban street three hours away in Burlington, a suburb near Toronto, Ontario, 15 years ago. Our journey to this place in the bush actually began over 30 years ago when we first considered the idea of recycling. We lived in an apartment in a medium-sized city, and I was looking at our trash one day and thought, a lot of energy has gone into making these cans and newspapers and glass that we're throwing out. Surely there has to be a way to do something with them.

We discovered a small group of like-minded people and began accumulating our recyclables and driving them to an industrial warehouse so they could be sold into the fledgling commercial recycling market. This was long before curbside pickup of recyclables.

A few years later we were living in our own home and the city had begun taking away our recyclables from the curbside. By then we had turned our attention to working to reduce our trash, getting it down to one can of garbage every eight weeks for our family of four. We were active in the local environmental group "The Conserver Society" and I served on the City of Burlington's "Sustainable Development Committee." Still, something seemed to be missing. We had a small house with no air conditioning and large black walnut trees that helped keep the house cool. Nothing grew under those trees in our backyard, so we had vegetables growing in the front yard.

I think it was after a visit to my chiropractor who asked, "Cam, was that corn I saw growing in your front yard?" that we decided it was time to get out of the city. The pace was too fast, there was too much of a focus on buying stuff and accumulating stuff, and as motorists got more stressed and switched to larger, wider, more dangerous SUVs, cycling was starting to feel just too risky.

So we established a five-year plan to move to the country. We put together a wish list for our dream home and started looking. After much time and many wasted trips we finally found it. We ran an electronic publishing business at the time and it seemed as though most of our customers would stick with us if we moved. Technology had allowed us to provide most of our services electronically, which made face-to-face meetings less necessary.

Moving three hours away from our customers was a scary proposition. Moving to an off-grid house with absolutely no clue about electricity was terrifying. Yet like many of life's defining moments, the greatest challenges presented the greatest opportunities. There were many hiccups along the way. It hasn't been smooth sailing, and there have been moments when Michelle and I have looked at each other and said. "Well, do we throw in the towel and move back to the city?" But they have been few and far between, and after we've taken a minute to re-evaluate the challenge of the moment we've figured out a strategy to deal with it.

Our learning curve has been huge. We went from not really knowing the difference between AC and DC electricity to doing workshops across the province that allow us to share what we've learned and inspire others to integrate renewable energy and sustainability strategies into their own lives.

We've learned about harvesting our own firewood, dealing with water issues in a rural home, and how much work is involved with growing much of your own food. There have been many successes and as many failures. We are now market gardeners and run a CSA or "Community Supported Agriculture." We grow food for families in our area who pay us to receive about 16 weeks worth of boxes of the organic vegetables we grow in our garden. The CSA has taught us a lot about growing food on a larger scale and just how much food we can get from the roughly one acre of land that we use for our gardens.

Michelle and I were contacted a few months ago about a new TV show about prepping. We went through a phone interview and then did a Skype interview with a casting agency in New York but alas, they didn't ask us to be on the show. This is probably not a bad thing.

I think the reason was we weren't chosen is that we aren't 'extreme' enough for TV. Apparently we are just way too mellow. Well, Michelle is anyway. Laid back apparently didn't work for them. Michelle and I ended up becoming independent because we wanted to lower our footprint on the planet, so we started producing our own electricity. Then we started growing more and more of our own food. So we basically fit the "prepper" profile from an independence point of view, but how we got there didn't seem to fit their requirement. If we had said we expect some rogue asteroid to hit Russia any day now (huh? is that even remotely possible?) and we're preparing for the complete disintegration of civil society shortly after, we probably would have had a better shot.

We are "accidental preppers". We didn't mean to become preppers, but suddenly we were.

The casting agency asked if we had any big projects planned, you know, the kind that make for great television, like building a concrete bunker, or fabricating a homemade grenade launcher from birch trees. I couldn't give them anything. After 15 years of fine tuning and basically replacing every system in the house, everything is finally working great here. We don't need anything else. I said I could tighten the guy wires on the wind turbine tower. You know, me with a socket wrench, looking frantic! And I've got to sharpen some chainsaw blades. Me at the sharpener, looking frantic! Anything, anything? Is that working for you guys?

We watched an episode of a show about a survialist building an off-grid house and he kept making these off the top exclamations like; "if I don't get this building complete before the snow flies, my family could freeze in the dark!" Then his wife would arrive from town in their minivan with a disposable coffee cup in her hand. Really? This is living on the edge?

The one overlying theme in our journey to independence has been what we tried to instill in our daughters while we home schooled them, and that's the concept of lifelong learning. Education does not end when you leave public school or complete your university or college degree—in fact that's just the beginning. Each of our challenges, whether it was adding more solar panels, putting up a wind turbine, or installing a solar thermal system to heat our hot water, has seemed complicated and often well beyond our skills at the beginning. But we read all the information we could find, consulted with everyone we could, and finally took the plunge and did it. We have had many great teachers along the way who've taken the time to share their knowledge and we are extremely grateful to them.

Now we want to share what we've learned. It comes from a lifetime of learning about how to be more independent. It comes from decades of research into the challenges that we face as a species and the steps you can take to prepare yourself for these changes. It also comes from a realization that this move to independence and accomplishing these tasks on your own creates an incredible sense of well-being and happiness. Years after putting up my wind turbine I still stand under it and watch it endlessly producing clean green renewable energy for my home and it feels as if my chest will explode with pride. It's way better than TV!

We have shared our journey to independence in our blog at www.cammather.com and in our book "Little House Off The Grid, Our Family's Journey to Self-Sufficiency." I also shared what I've learned in 30 years of growing vegetables organically in "The All You Can Eat Gardening Handbook."

This latest book, the book that you now hold in your hands, grew from my 2009 book "Thriving During Challenging Times, The Energy, Food and Financial Independence Handbook." I have updated and refined the information from that book. In the wake of the devastating weather events that seem to be hitting every part of the country from the coastlines to the interior, I have put a greater emphasis on emergency preparedness. This seems more relevant than ever. I have also updated the information on food and water storage. This summer's drought was the worst since the Great Depression. I was growing food for 12 families and we did not have rain for 8 weeks during the prime growing season, yet our customers raved about the volume and quality of the food. If I hadn't planned for such a situation I would have had angry customers who would be unlikely to join in subsequent summers.

Michelle and I are writing this book to share our experiences and to try and help you avoid more anguish than is necessary in a changing world. We have experienced great challenges in our move to self-sufficiency, but we chose this path and the pain was incremental. When I see the disruption of so many caused by the radical climate events happening today my heart really goes out to those affected. We have a little experience in this department. Our farm includes a guesthouse that was built by the previous owners. It houses two bedrooms and my office upstairs, and garage and a storage area for the books that we sell downstairs. The battery room where all of our renewable energy equipment is concentrated is also located on the ground floor. Unfortunately the structure was built at the bottom of a gully. When the original homesteaders built our house in 1888 they built it on a high spot. Twenty years ago some contractor working on our garage apparently didn't notice that all the ground nearby ran down towards the building.

Most of the time this is no problem. In big rainstorms the water is absorbed by our sandy soil or runs off towards our pond. But about once a year we have a flood. Usually in March we will get a thaw accompanied by a big rain. Because the ground is frozen it can't handle the water so it all runs down to the guesthouse. I've tried to landscape around the building but regardless of how much I try and keep the channels clear, some ice dam will form while I'm asleep and the building will flood. And it's devastating. Water seeps in between the concrete slab and the wooden walls. Drywall gets wet. Insulation gets wet. And having 2 inches of water slopping around in a room with the batteries and equipment that powers our life is really distressing. I've figured out how to clean it up quickly and I crank up the woodstove to get things drying out, but when I walk in and see the water in that building, I feel like sitting down and having a cry. Sorry, but it's true. I know, it's only "stuff,", but it is very distressing regardless.

So I think about the anguish I feel when my guesthouse floods, and then I ratchet it up about 1,000 times to get a sense of what it must feel like to see your entire house underwater. Or to see your house in your neighbor's yard, after a tornado. It is heart breaking to watch. We can't control the path of a tornado but we can take some measures to try and mitigate the impacts of these events.

What you need to do now is to start creating an independent living space to ride out any storms that may be coming. The technology exists to keep you living a pretty comfortable life even if the electrical utility is having trouble keeping the power on. At our house there are no power lines or utility poles of any type bringing energy to our home. We are almost completely independent in terms of our energy, food, and income. If need be, we could be completely independent and our lifestyle would not notice a real decline. This is the goal you should be striving for, and the technology and knowledge are here, today. They're in this book!

In a recent radio interview I suggested to the interviewer that I could close the gate at the end of my driveway, and not leave or have anyone deliver anything for 6 months, and my life would go on pretty much as normal. I would probably run out of coffee and cream, which would make me cranky for a while, but I'd get over it. Then I'd get right back to being warm, to having running water, lights, radio, TV, internet, a refrigerator, a freezer and just about anything a typical North American household would have. The only difference is that I'd be producing the energy myself. The water would come from my well and travel to my faucet using the power I generated from the sun and wind. And I would eat very well. I would miss luxuries, but I'd be comfortable and content. Now imagine you not leaving your house for 6 months. How long would you live comfortably? And nothing could come in either. Ordering everything for delivery to your house would be cheating.

There's never been a better time to get started on your own journey of discovery about the challenges we face and the steps you can take to prepare. It can be a wild ride, but it's a blast!

3 The Challenges

Economic Challenges

In 2008 the U.S. economy was sitting on a ledge overlooking a precipice. Millions were losing their jobs. Millions were underwater on their mortgage payments. The big banks required a $700 billion bailout. Lehman Brothers went bankrupt. AIG Insurance required more than $100 billion for its bailout. GM and Chrysler were bankrupt. Fanny Mae and Freddy Mac had to be bailed out. Things were not good. Some called it "the great recession" although many economists felt the depth of economic turmoil was comparable to the Great Depression.

In the time since then, the economy appears to be better but the housing market is a shadow of its former self. Millions still remain out of work. It took a lot of money to fix things, which the government created and took on as debt. It just kept piling more debt on its already big debt. In the previous depression many economists felt that the government had made a mistake contracting their spending, so this time they went out of their way to open the spending floodgates.

Printing money to spend your way out of an economic collapse may seem like a good idea at the time, especially if you plan on curtailing spending once the economy recovers. Unfortunately governments are always good at spending but have trouble cutting their spending later. In 2012 the U.S. Federal Government spent about $1 trillion more than it took in. It's done this for the last 4 years. The accumulated government debt is more than $18 trillion. It goes up by almost $4 billion a day. Each American's share of the debt is around $56,000. So not only do you have personal debt, the Federal Government has accumulated some debt on your behalf.

There was a time when government debt was thought to be acceptable because governments could grow the economy and that would help pay off the debt. In a world as economically challenging as ours, it's difficult to see that level of economic activity as realistic. In the past some governments have inflated the debt away by rising interest rates, but governments seem determined to keep interest rates low to get the economy on track. So this doesn't look like a solution.

The $16 trillion debt is not the whole picture. The government has "unfunded liabilities" which are obligations it has to pay which it hasn't set aside money for. These are things like Medicare, Medicaid and Social Security. When George W. Bush was president he created a national prescription plan for seniors which added trillions more to unfunded liabilities.

On May 28, 2008 Richard W. Fisher, President and CEO of the Dallas Federal Reserve Bank, estimated the obligations of the U.S. taxpayer to be an impossible-to-repay sum of $105,200,000,000,000.00 ($105 Trillion).

Economic historian Niall Ferguson has calculated that total U.S. unfunded liabilities are more than $200 trillion.

The numbers are starting to get to the point where it seems that there may be no way for the government to actually make good on all of these obligations. Throughout history governments throughout the world have defaulted, or basically declared bankruptcy. We haven't heard about it too much lately but it's not a pretty sight. If the U.S. reaches this stage things will get very bad for most North Americans. Many people rely on the government directly or indirectly. You don't have to be a public servant. You may work for a steel company that supplies product to the U.S. Navy. Or you may sell cleaning supplies to the steel company that supplies the Navy. Suffice to say if the Federal Government is unable to pay its bills, the repercussions will be far reaching.

Right now things seem to be going all right in the North American economy and that's a good thing. We can only hope that they continue. But sometimes reality can have a way of sneaking up on us and causing drastic changes. So keep this in mind as you start to formulate your "Plan B".

Peak Oil and Resource Depletion

The world was granted a one-time inventory of fossil fuels like coal, natural gas and oil. They were created over millions of years as organic matter built up in successive layers and under heat and pressure carbon-based fuels were formed. Millions of years later humans learned we could burn them. We started with wood, then switched to coal and then discovered the wonders of oil and natural gas.

In the 1970s the United States hit "peak oil." This was the point in time where the U.S. had extracted half of its one time allotment of oil from the ground. There was a bit more discovered in Alaska and good supplies in the Gulf of Mexico later, but if you look at the graphic on the next page you can see that U.S. oil production went into decline and has never recovered to the same levels as the peak. Currently new technologies such as hydraulic fracturing of rock, or "fracking" are allowing oil trapped in rock formations to be freed up. Under great pressure water, sand and chemicals are pumped into these rock formations to fracture it and free the oil trapped in smaller pockets. Fracking uses a lot of energy and adversely affects local water supplies. The U.S. Geological Survey has suggested that fracking also leads to seismic or earthquake activity in areas where it is undertaken.

In the past when you drilled for oil or natural gas, it would shoot out of the ground under pressure for you to capture. In those early wells for every unit of energy you put in you might get about 100 units of energy back out in the form of oil. Now this "energy returned on energy invested" ratio, or EROEI, is getting smaller and smaller. We use more and more energy to get that oil or natural gas out of the ground. The tar sands in Alberta Canada require monstrous excavators to load building-sized trucks with the tar-like "bitumen." This bitumen is transported to facilities that use massive amounts of natural gas energy to heat, bubble and extract the low quality oil from the sand. In this case you might have an EROEI of 3 units of energy out for every unit of energy you put in. So we have to use way more energy to get energy. And when you reach the point where for every unit of energy you put in, you only get one out, it's not worth doing anymore. You're just trading one form of energy, like natural gas, for another, like oil.

So while we are discovering new sources of energy like oil and natural gas in the Bakken shale of North Dakota, or deep in the waters of the Gulf of Mexico, it is much harder to extract. We're using more energy and capital to get it out of the ground. And oil is going to continue to be more expensive to purchase.

Since 2005 the world has been on a plateau of about 85 million barrels of oil extracted from the ground each day. During this time oil has been as high as $148/barrel and over the last few years has remained close to $100/barrel. This is a huge run up in price since a decade ago when oil was priced at around $20 or $30/barrel. If you use a classical approach

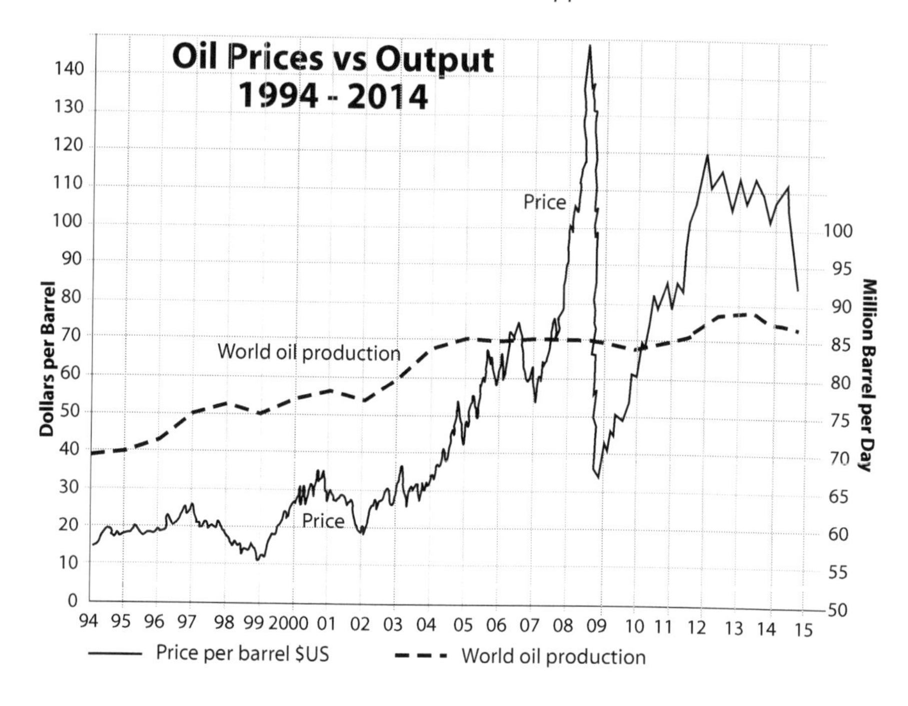

to economics to look at the situation you would expect that the higher prices would have inspired more activity to drill for oil and bring it to market. When the prices of a product rise, supply generally increases. And yet when you look at the price line of oil overlaid over its production it seems we have hit a plateau. Higher prices are not increasing supply which would drive prices back down.

Many petroleum geologists suggest that the plateau represents the fact that the entire world has now hit peak oil. The world has extracted half of that one time allotment and soon will begin to decline. The higher price is inspiring some new supply to come online but it's not enough to stop the decline, it's merely replacing what's being used up. The year we discovered the most oil on the planet was 1964 and since then we've been discovering less and less each year. Previously once in a while we'd discover a huge oil field or an "elephant" but today the pockets we find are smaller and smaller. Or they are in extremely difficult locations to extract it from.

The International Energy Agency (IEA) which advises many world government about energy issues says the world hit 'peak conventional oil' around 2005. This is the easy oil to get at, where you drill a well and it comes out under it's own pressure. We have not hit the peak of non-conventational oil such as shale oil or tar sands that takes way more energy to get out of the ground.

Remember the 2010 explosion and fire and oil spill into the Gulf of Mexico from BP's Deepwater Horizon drilling rig? This was a prime example of the hazards we face using current drilling methods. We are drilling for oil deeper and deeper and often times we lack the technology or human capability to do it safely. The oil spill in the Gulf was a wake-up call to the world that the days of cheap, easy and plentiful oil are over. Oil is going to get harder to extract and more expensive.

So why is this such a big deal? Oil plays a huge part in our lives and allows our society to accomplish incredible things using this "energy slave." Having access to oil is like every North American having a staff of servants to do much of the tough, physical work for us. Oil provides us with the equivalent of 300 slaves working for us each and every day. And our energy slaves don't require us to feed them. Oil just keeps doing hard work at an extremely low price.

Our economy is also based on great quantities of cheap energy. So as the price of energy rises the economy suffers. Some economists suggest that the reason for the economic collapse of 2008 was more than just financial malfeasance on Wall Street. They suggest that with oil approaching $150/barrel the economy could no longer function as it always had with lots of cheap oil driving it.

At the same time as we seem to be hitting a plateau of energy production and may start to go into decline, there a number of economies that are just starting to use oil in large quantities. China and India have huge populations just starting to achieve incomes that allow them to consider car ownership and they want their piece of the pie. So as energy supply is starting to decline, demand is starting to increase. This is not a good thing for political and economic stability.

Many of us would like to think that perhaps solar and wind power could help replace a lot of this fossil fuel energy that is getting more expensive, but the reality is that it is very difficult to duplicate the tremendous energy contained in fossil fuels. Living off the electricity grid I encounter this reality everyday. I have a propane stove, but I cook with electricity I generate from my solar panels as much as I can. When I see how many solar panels I require to do this and how much of my overall energy supply I have to devote to creating heat to cook with, it's quite unbelievable. When I turn on the burner on my propane stove I marvel at the ease with which I get good quality heat.

Anyone who has ever pushed a car very far knows the wonders of fossil fuels. It's tough to move a car very far even if it's a compact car. It's amazing how quickly you become winded. If it's a pickup truck I think you'd be lucky if you could even get it moving at all. So now think of how much energy would be required to get that big hunk of steel and rubber up to 50 miles/hour. Imagine pushing it up a big hill at 50 miles/hour. The embedded energy in a small quantity of gas and oil is truly mind-boggling.

The message is that the future is going to be quite different from the past. Energy is going to be more expensive. The economy will continue to struggle as high energy costs restrict its ability to thrive. And shortages are a possibility. After hurricane Sandy we saw huge lines of vehicles at gas stations. People were spending the better part of their day just trying to get fuel. The shortages were widespread in the region. So you need to resolve to not be that person in line hoping to get some gas. You need to have your own "strategic petroleum reserve" for when the lines come to a gas station near you.

Peak Food

There is a group called the "Club of Rome." It sounds like either a hip European dance club or a sinister, elite old boys' club. It is in fact a global think tank that published a report in 1972 called *Limits to Growth*. The report suggested that the world population, which had reached 3 billion in 1961, was at a theoretical threshold and that the planet couldn't support any more people without seriously damaging its ability to support life. The report sold 30 million copies and was a topic of conversation at the time, but the population continued to grow exponentially.

Improvements in technology, the bioengineering of crops, and the eventual lowering of oil and natural gas prices, which allowed lots of natural gas to be converted into fertilizer, have dramatically increased crop yields and populations have continued to grow.

But there are cracks starting to show in the theory of infinite growth as evidenced by the simple fact that the population of the planet doubled in just four decades, from three billion in 1960 to six billion in the year 2000. We've now hit 7 billion. That is a lot of people in a very short period of time. That's a lot of mouths to feed. If you look at the population of the planet over time, it grew very slowly for many centuries. Then we unlocked the fossil fuel genie, and suddenly farmers could grow huge surpluses of food to feed more mouths.

Basic logic tells us that this simply can't go on forever. Sooner or later we will hit the limit on how many people we can feed. When you look at how explosive population growth has been in the last 50 years, most people can make the leap and acknowledge that we've hit that limit or perhaps gone beyond it.

North Americans basically eat fossil fuels. For every calorie of food energy we eat, 10 calories of fossil fuel have been used getting it to our plates. This includes;

- the natural gas for fertilizer
- petroleum products for pesticides and herbicides
- diesel for the tractor to plant the seeds, apply chemicals and to harvest the crops
- diesel to truck it to a processing plant
- energy at the plant to pump water to clean it or to boil water for cleaning or canning it
- energy to make the glass or steel for the container that the food ends up in
- energy to make the cardboard for the box they are packed in
- diesel for the truck to get it to the grocery store chain's central warehouse
- energy to keep that building cold
- diesel to take it from the warehouse to the food store
- electricity to keep the lights on and refrigerators cold at the store, and finally
- gasoline for you to drive to the store and buy the food.

In the past as we've run into the limits of our ability to feed all the new mouths on the planet we've come up with some innovation to avoid disaster. First we harnessed fossil fuels to allow us to get more food out of our fields. Then we came up ways to crossbreed and hybridize plants to produce more, and now we are trying to genetically modify plants and animals to produce more from less.

Many experts are now suggesting that we're starting to run into theoretical limits on just how much food the planet can produce. This comes at a time when not only is the population still growing by 70 million each year, but many people in developing economies are starting to have more wealth to spend on food and that often includes an increase in animal protein. Animals require more land and resources to produce a given amount of food than someone just eating plants like grains and rice, so this put an even greater strain on the food supply.

As food shortages start to occur around the world governments often become less willing to share their resources in the form of exporting excesses. They want to keep some in the pantry for their own populations. So while North America produces a lot of calories for everyone who lives here, as world food prices rise so will ours. Food will get more expensive and we could start experiencing shortages. Having a plan to feed your family makes more sense than ever today.

Peak Water

All forms of life need water. Even single-celled animals and plants like bacteria and algae need water to survive and function. Living organisms are made up of approximately 60 to 70% water by weight. Water covers about 75% of the surface of the earth. The amount of water on the earth remains fairly constant, as it is continually "recycled" through evaporation and precipitation. This cycling of water plays a role in regulating the climate of the earth.

Less than 3% of the earth's water is fresh water and most of this fresh water is frozen in the ice caps and glaciers. Fresh water is a critical part of our world that is in relatively short supply!

The International Energy Agency notes: "The Earth has a finite supply of fresh water, stored in aquifers, surface waters and the atmosphere. Sometimes oceans are mistaken for available water, but the amount of energy needed to convert saline water to potable water is prohibitive today, explaining why only a very small fraction of the world's water supply derives from desalination."

The head of the Food and Agricultural Organization of the United Nations (FAO) says two-thirds of the world's population could be threatened by water shortages by 2025. The World Bank reports that 80 countries now have water shortages that threaten health and economies while 40% of the world's population—more than 2 billion people—has no access to clean water or sanitation. In this context, we cannot expect water conflicts to always be amenably resolved.

Fortune magazine recently said; "Water promises to be to the 21st century what oil was to the 20th century: the precious commodity that determines the wealth of nations." Water, like oil, is going to become increasingly important on the world stage. In North America we're already seeing water shortages starting to have a negative impact on many people's standard of living.

So you need water and you need a strategy to ensure you and your family has a good supply of it ready to use should there ever be a shortage.

Climate Change

I believe 2012 was a turning point for many people regarding the reality of the climate changing in a negative way. The United States experienced record wildfires. Tornadoes were ferocious, and the drought that gripped much of the country was epic. Even during the winter shipping on the Mississippi was in question because water levels were so low and the drought continued. Then came Superstorm Sandy with its unparalleled size and destruction. Along with its horrific winds and rain came a storm surge that sent water into areas that had never experienced flooding before. The surge was compounded by the fact that the ocean is rising as it gets warmer and more polar ice melts.

Large areas of New York and New Jersey suffered power outages and tremendous damage to homes and infrastructure. Millions of people in the richest city in the most powerful country on the planet were without power, heat, food and water. Parts of the New York subway flooded by seawater will take millions of dollars and months to repair. And the $60 billion that is the early estimated cost seems low when you look at the number of damaged homes and public assets from roads to boardwalks.

When Hurricane Katrina devastated New Orleans it took the U.S. Government 10 days to release billions of dollars of relief funds. When Hurricane Irene devastated the Northern States it took Congress 12 days to release funds. Seventy days after Superstorm Sandy no funds had yet been released. Part of this delay was because of the "Fiscal Cliff" that was preoccupying Congress but there seemed to be less urgency at higher levels to move quickly on aid. It seemed like a bit of disaster relief fatigue was setting in.

And it's no surprise. In the 1990s the International Monetary Fund reported that natural disaster destruction averaged about $20 billion each year, but between 2000 and 2010 it had increased to $100 billion. Superstorm Sandy will set a new high threshold. While the IMFs figures are worldwide, the bulk of this damage is the United States. This is partially because with its wealth it has more infrastructure to be damaged in each storm. It also seems to be uniquely situated in terms of hurricanes coming across the Atlantic and has a number of other areas that are prone to extreme weather events.

Regardless of whether or not you support the concept of anthropogenic or human-caused climate change, our new reality seems to include rising oceans, more unprecedented rains and floods in some areas, droughts effecting other areas, and increased storm intensity and rising damage costs. Having a plan on how your family would respond to such an event is critical.

Complexity and Resilience

There were advanced societies before we came along, such as the Mayans and the Romans. People living in those societies would have thought of their times as extremely advanced. They had great diversity of labor. There were farmers and hunters and merchants and trades people who made things. There was also a group of people who were the rulers. There were people to collect taxes to support the work of this government structure, but these cultures were advanced enough that they were able to produce excess food to support this structure of people who didn't actually "do" anything. They governed. Or they were spiritual leaders. But they didn't grow food and they didn't produce anything tangible.

Many empires made great use of slave labor to accomplish their dominance of regions. Today many people feel the United States with its huge military might and far reaching political clout very much resembles an empire. Today the U.S. no longer uses human slaves but has captured the power of energy slaves through fossil fuels. It's an empire that very much runs on energy slaves.

Many refer to the Mayan and Roman periods as "Empires." As these societies grew they controlled larger and larger areas and more and more people. Part of this growth was based on the fact that as the ranks of unproductive people grew they required more and more people who made "stuff" and grew "food," to supply everyone else. And as the people in charge, or the governments, got bigger they grew more complex. They simply had larger areas and more people to administer. And part of that complexity came from having to keep the people they governed under control. They needed armies to do this, and again, technically the army didn't do anything directly. Not only did they not grow their own food, more food had to be grown by everyone else. So the empires continued to grow and spread to try and incorporate more farmers to support the armies.

You can imagine how this becomes a vicious circle. Your society feeds upon itself. You need to encompass a larger area to have more food and energy inputs to support the growing number of unproductive people (army and government) but as you grow the area you control, you have to employ more people to keep all those you control in line. You need a bigger army. You need more slaves. You need more tax collectors. You need more government.

As societies grow in size they therefore grow in complexity. There is greater infrastructure required to maintain the empire. As you add more layers of administration and bureaucracy things just get more complex. As complexity grows the empire becomes less resilient. It becomes more vulnerable to a shock which can disrupt the whole system. As empires start having problems they deal with them by adding more layers of complexity. More complexity equals less resilience.

Many anthropologists have documented these concepts over the years. Joseph Tainter looked at this concept in his book, "The Collapse of Complex Societies." Tainter suggested that as societies grow more complex one of the layers of complexity they add is a class of information producers and analysts. In the Roman Empire they would have been people who administered the many regions under Roman control. If an emperor wanted more income he could turn to them to decide if a given region could support higher taxation. Or they could tell him whether a given region could produce more food through improved agriculture and irrigation, which could support more Roman troops of occupation in the region.

Now think of our society today. Think of how many people would be considered information producers and analysts. Government workers at all levels; federal, state, provincial and municipal. People who work on computers. Lawyers. Accountants. Writers. Journalists. Then think of all of the other professions not involved with actually making stuff. Teaching at all levels from public school to colleges and universities. The medical profession. People who work in retail selling stuff. In fact as much as 70% of the U.S. economy is based on buying stuff, much of which is made in other countries. Today less than 20% of the economy is actually involved with producing "stuff." And farming, well those numbers are horrifying. At the turn of the last century, in 1900, a farmer grew enough food for about 7 people. Today a farmer is feeding more than120 people. That's an awful lot of people depending on every farmer. I don't how they sleep at night with that kind of pressure. When they plant and harvest those crops the tractor is often being driven automatically by a satellite controlled GPS system. Talk about complexity.

According to the EPA in the last census there are only about one million Americans out of a population of more than 300 million that claim farming as their principal occupation. So less than 1% of the population is growing food for the 99%. In 1890 a human worker could work 27.5 acres of land. In 1990 one worker could look after 740 acres. I would suggest that in the more than 20 years that have passed since then, that number will have again increased significantly. Farmers are able to grow on this scale because of fossil fuels used in their tractors and in the insecticides and pesticides and herbicides made from petroleum. This reality of our dependence on such a small number of people to feed everyone else is truly disturbing.

So North American society has basically become one of information processors. Economists even hail this as a great accomplishment. Sure we've lost those high paying manufacturing jobs that you didn't need a college education for, but we have all these great new information jobs, high tech jobs, jobs where we push pixels around a screen. As our society has transformed into this information based economy we have made ourselves much more vulnerable to shocks. We now rely on other countries for many of the goods we consume. We rely on other countries for much of our fossil fuel energy, certainly oil for transportation. We rely on fewer and fewer people with bigger and bigger and more complex farms to grow our food.

Every time we've added another layer of complexity or "farmed out" an essential component of our lives, we've made ourselves more vulnerable to a shock. A drought. A shortage of cheap energy. Political upheaval in another part of the world. Now our closely interconnected economy has made us much less resilient and much more exposed to shock.

Collapse is unlikely to come overnight. Humans are resourceful and we find ways around challenges. But empires that came before ours succumbed over time. Sometimes over decades. Sometimes over centuries. But ultimately their societies grew too complex and they collapsed. When the Roman Empire collapsed Europe entered "The Dark Ages." While the Roman Empire was noted for its brutality and use of slaves, historically many view it as a time of great progress for civilization. After its fall the world experienced depopulation and deurbanization, as people who had lived in cities were forced to move back to the land to grow food. Society became less complex. There were fewer layers of complexity. Most people just grew food and supported themselves.

The United States spends more on its military than do the countries with the next 13 highest budgets, combined. In 2011 it was more than

$700 billion. Its military uses a mind-boggling amount of energy. It has 11 aircraft carrier groups. The country some see as posing the greatest military threat in the future, China, has just purchased its first aircraft carrier, used, from Russia. The U.S. has hundreds of military bases in countries throughout the world and more than 2 million people working on them. It's a big organization. It sure looks like an empire to most people.

So the U.S. today is hugely complex with many, many layers of complexity, and it requires a huge amount of energy to keep it functioning. It looks like it has much vulnerability, not to foreign attack or invasion, but from collapse within.

Why it's different this time

In the past advanced countries have been able to pool huge resource bases and tackle seemingly insurmountable challenges. When the United States entered World War II it took the entire resources and commitment of the country to win victory. Everyone pitched in and did without to help the cause. When the U.S. decided to put a man on the moon it took an enormous amount of intellectual and financial resources to accomplish something many people felt wasn't even possible. In the past when faced with a challenge the United States has risen to the occasion.

If our society faced one such challenge today, we could probably do it again. If we just had to deal with the effect of peak oil I'm sure we could make an all out investment in new technology and infrastructure that would help us to adjust to a new energy reality. If the U.S. decided that it had to take dramatic action on climate change to stave off disaster down the road, and it threw its entire and formidable resources into the fight, I have no doubt that the world would follow suit and we'd see a dramatic decline in greenhouse gases entering the atmosphere.

And yet today we are faced with a myriad of major challenges. Each one is big and has many causes. The reality is that today it seems there are just too many people on the planet using too many resources, too quickly. The earth only has so much and we seem to have crossed the point where it can provide anymore. So governments today aren't just faced with resource depletion and peak oil, they are faced with climate change, and water shortages, and food shortages and a huge economic dislocation that started in 2008. All at a time when most governments are heavily in debt and don't really have the resources to deal with any of the problems well. Governments don't have 'rainy day funds' they can tap into to help. In fact many governments are beginning to speak about "austerity," of spending less, paying down debt and trying to get their financial houses in order.

What this means increasingly is that you're on your own. Governments may allocate some money to help you out in a crisis, and your neighbors will try and do what they can, but basically we're all in the same boat. So you need to increase your own personal resilience to the shocks our system is facing. Most of us have the resources to increase our independence but inertia just keeps us spending as we always have. There's always a new electronic gadget that looks pretty cool, and those trips south in January look so inviting, really how could we pass them up?

Well you can and you should. You need to reprioritize how you spend your money with your first question always being, 'if I spend this money will it make me more resilient?' Now let me give you the tools for making those smart choices.

4 Starting Your Plan

The future isn't going to look like the past

Which means it's time for you to stop thinking it will. That retirement to the beach or golf course with ample funds and plentiful and cheap energy to make your life easy may not be going to happen. In fact, for many who have watched their retirement funds drop precipitously in the financial meltdown, this reality is already hitting home. Some might argue that the value of those retirement plans was artificially inflated during the last 20 years without the underlying economic strength to sustain it. If that's the case, it was easy come, and now it's easy go. That's not an acceptable or particularly comforting explanation to most people, but it may be the new reality.

In the depression of the 1930s there were virtually no government entitlement programs. There was no welfare. There was no unemployment insurance. There was no government-sponsored health care. There was no old age security or retirement benefits. People were simply more independent. People didn't expect the government to help them out when things got tough because it had never done so in the past. When things got tough, people had to get going to get themselves out of their situation.

We seem to be moving into a period of dramatic shocks to the systems that we've developed as humans. Shocks to our economic system, shocks to the climate we live in, and shocks to the energy systems that provide North Americans with such a high standard of living. So I would suggest it's time you started to make yourself more resilient to shocks.

The following chapters are a road map to guide you on your way. They look at all the major "needs" that most humans have. Many of these are located on the lower rungs of Maslow's hierarchy of needs: the need to eat, the need to stay warm, and the need to stay safe. The future will require many of us to spend much more time and energy dealing with

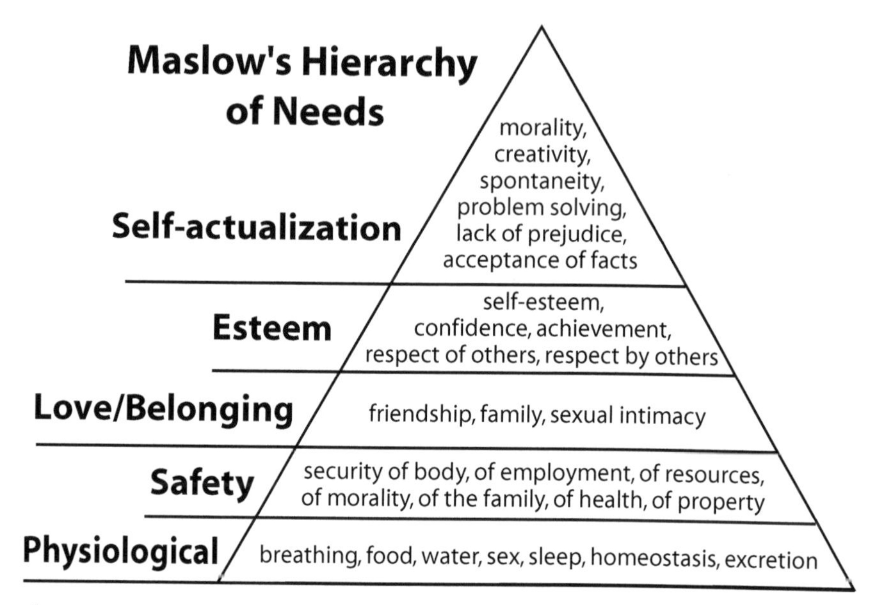

these issues. But this is not to say that you won't also be involved with the higher rungs at the same time. In fact, you may find that dealing with basic human needs brings a new joy and sense of fulfillment to your life and that those higher-level needs start looking after themselves.

Spending a lot of time commuting to a distant job and working horrendous hours may have been affecting your ability to enjoy as positive a relationship with friends and family as you would like. With less income you'll have more time to spend growing your own food and enjoying other people's company, both of which are free.

While many of us have defined ourselves through our jobs, we have often felt underappreciated and ignored in the corporate world. Superiors haven't respected us or listened to our ideas, which can affect our self-esteem. Producing your own electricity with renewable energy and growing your own food is incredibly invigorating. There is nothing like the feeling of accomplishment you get from taking that first basket of vegetables out of the garden or watching your solar monitor tell you how much electricity your solar panels are producing This can be incredibly freeing, invigorating … and good for the soul.

In fact every suggestion that follows is a good thing on a number of levels.

First, these ideas will help you and your family deal with the increasing shocks to the system that are going to accelerate in the future. The suggestions that follow will make you less vulnerable to energy, food, and financial shocks. They will make you more resilient to these crises.

Second, they will be good for your health, both mentally and physically. Riding your bike more often is good for you. Growing your own vegetables is excellent cardiovascular exercise and, as so many people have discovered with flower gardens, it is an incredibly satisfying and enjoyable activity. Hoeing weeds lets you "zone out" and reflect on things that make you happy.

And finally, everything that follows is good for the planet. It will help you reduce your footprint on the planet, and the planet desperately needs you to do this, now! Using renewable energy to make yourself more energy independent also helps you reduce how much coal is burned to make the electricity you use in your home. Eating food grown in your garden or by a local farmer dramatically reduces the energy required to get food to your plate and therefore lowers the carbon produced. Harvesting your rainwater rather than having it run down your driveway into a city sewer reduces the stress on your municipal infrastructure and helps to eliminate raw sewage being washed into local watercourses when the rainwater sewers become overwhelmed during storms.

I would suggest you start developing a "Five-Year Plan." Take the information provided here and chart a course for your next five years. Have a goal, whether it's to buy a smaller house closer to work so you can take transit or to buy a place in the country so that you can grow most of your own food. Figure out what it's going to take to get there, and then develop a plan. Come up with realistic financial goals and timelines, and then put them on paper. Keep referring to the plan. Put it on the fridge. Keep checking it. Put it on the breakfast table next time you're considering a big purchase. Do you really need a new couch or TV, or will the existing one make it through another year? If it will, put that $1,000 right into the "Five-Year Plan Account." You won't be so disappointed about watching movies with those black bands at the top and bottom of your "non-widescreen" TV anymore. Those black bands represent that house in the woods you've always fantasized about. That $1,000 you didn't spend just brought you that much closer to the goal.

Michelle and I have chosen to structure our recommendations in two ways. First we are going to provide tips for dealing with the most pressing of challenges you may face… the "clear and present dangers." These could be an approaching tornado or flood, or an extended power outage that seem to occur with increasing regularity. These will be relatively inexpensive suggestions that we'd suggest you make in the short term.

Tomorrow would be a good day to start. Today would be even better. Fumbling around in the dark looking for a flashlight is not the best way to start dealing with a crisis.

Secondly we're going to offer longer-term tips and suggestions on how to look a little further down the road with your prepping plans. It may be years before any great disturbance comes into your life. And that's a good thing. But there is a strong likelihood that the challenges we face are going to affect you in some way, so you should start to make more significant plans. You need to think about things like where to live. If you live close to the coast right now you have to decide if that's the best long-term plan. If the oceans are rising and storm surges are getting more intense, does it make sense to be living at sea level in a potential flood plain? And if oil is going to get dramatically more expensive how realistic is your hour-long commute in a pickup truck? Is there anything you can do now to start moving away from being vulnerable to potential shocks to the system?

We're going to provide lots of tips and strategies. You will find some useful and others won't be applicable to you and that's fine. Find what works for you. But please start taking some action. The urgency of improving your family's resilience to shocks increases each year. It increases each season, as the storms get more intense. It increases each quarter as the economy groans to move forward. It increases each month, as the electricity poles in your neighborhood get older and less able to handle those windstorms.

So, let's get crackin'!

PART II

Clear and Present Danger - Emergency Preparedness

5
Warning Signs and Running for Cover

Extreme weather seems to be the norm, so we should all know how to recognize the danger and be prepared to respond to it.

Some natural events, like earthquakes are difficult to anticipate. Warning signs, such as intense thunderstorms, often precede others, like tornadoes. So this chapter is a quick recap of some things you should be aware of.

Emergency preparedness organizations like FEMA (www.fema.gov) and the Government of Canada (www.getprepared.gc.ca) list various hazards and potential emergencies. Some you can control and anticipate and others you can't. So here are some general thoughts about the common hazards.

Low-level dangers

Drought and extreme heat were the norm in the summer of 2012 which was the hottest year on record for North America. There isn't much you can do other than follow some of the tips I suggest in Section Three in terms of water storage and building an efficient home to help deal with heat.

The flip side of this is extreme cold like we've experienced with the polar vortex. One thing scientists have identified about climate change is that weather "extremes" are becoming more common. Most of the extremes are heat related, but periodically you may experience periods of extraordinary cold weather. While it's easy to get complacent as winters get warmer you should have an adequate backup plan for cold weather as well. Assume the energy supply lines that keep your home warm, natural gas, electricity, oil, propane, may become disrupted. You need to have a backup heating system which I discuss in detail in Section III.

Wildfires are in this same category. If you are experiencing a drought there is little you can do to prepare for approaching wildfires. The best preparation in this case is to minimize the number of trees and bushes that are planted close to the house. These could become dry and stressed and easily catch fire from sparks from the wildfire, which could then spread to your home.

I would categorize volcanoes as a low level danger too. If you lived near Mt. St. Helens before it erupted you would have been advised to know the risks and have left the area soon after the eruption started.

Yellowstone National Park is located in a "caldera" which is basically a volcanic crater. It is surrounded by what was once the rim of a volcano. Its hot springs and geysers come from the magma that located near the surface. The volcano there erupts once every 600,000 years or so and some geologists feel that an eruption may be eminent. This is partially based on seismic activity in the area, as the number of earthquakes greatly increased in recent years. If it does erupt it will be a big deal, called a super volcano because of its size. The odds of this are low, but not zero. So you could build a house nearby and never have a problem, but some would suggest that wouldn't be prudent. The reality of a volcano that big is that it would also have an effect on a huge part of North America and indeed the world as massive amounts of volcanic ash would be blasted into the atmosphere and severely restrict sunlight hitting the earth. This is bad for people who grow food and for people who rely on renewable energy like solar power. In this case I'm in a bad way because I do both! I've got my fingers crossed about another big volcanic event.

I must also mention a danger which has a fairly low probability but a high potential impact, and that's a coronal mass ejection. The sun, which is a burning mass of hot plasma, emits all sorts of interesting things like radiation that keeps us warm and various magnetic storms and things. Periodically the sun will have a massive burst of solar wind and magnetic radiation called a coronal mass ejection. We know these have happened before and that they can greatly disrupt our electronic communications equipment and basically anything electronic. Fifty or a hundred years ago this would not have been a big deal, but today in our wired world where everything from phone networks to power grids are susceptible to such a magnetic pulse and every device we own seems to have a computer controlling it, such an occurrence would have a profoundly negative impact on our society. Again, the likelihood of such an event is low, but the potential negative impact is extremely high.

Moderate level dangers

Avalanches and landslides are not something most of us have to worry about. It's a concern for people who live in beautiful, mountainous regions. I guess there has to be a downside to living amongst all of that natural beauty.

People who live in these areas are generally aware of the dangers. If you ski the backcountry or snowmobile in the mountains the avalanche danger is usually well publicized so you need to have this information and respond to it. Every year many people are killed in avalanches. While it's not a way I'd want to go, it certainly is a risk many people feel comfortable taking. If you are going to be in these dangerous areas make sure you have an avalanche transceiver or beacon which will assist your rescuers and improve how quickly they are able to find you and increase your chances of survival.

Landslides are like avalanches but without the snow and again, happen in mountainous areas usually during periods of extensive rain. If you live in one of these areas know the risks. Have an idea what surrounds your house and be aware if the soil is prone to becoming water logged and might eventually giving way. And if in doubt, during torrential downpours stay clear until you can have a qualified expert confirm that the area is stable.

I always heard that my chances of winning the lottery are lower than getting struck by lighting. About 50 people are killed in the U.S. every year by lightning strikes. Many more are struck and injured. I remember being on a long bike ride once when a really bad thunderstorm struck so I did what many people instinctively do, I took shelter under a tree. Now think about it. Lightning is looking for a path to the ground. Something high and easy to hit. So how logical is it to stand under a tree? You might as well take a long metal pole and stand on a hill!

If a thunderstorm is approaching and there is lightning, get inside. Even a car is better than nothing, just try to avoid touching any metal. If you're in your house unplug electrical devices like computers and TVs. Don't take a shower or do a laundry because the metal pipes may conduct a lightning strike. Stay away from doors and windows and stay off the front porch. Stay away from anything metal like a motorcycle or bicycle or tractor. And if you are outside, avoid hilltops, open fields, the beach or a boat on the water. These will just make you a target.

If you were to find yourself in a situation where you could not find shelter in a lightning storm here's a recommendation from the New York

State Department of Health; "If your skin tingles or your hair stands on end, a lightning strike may be about to happen. Crouch down on the balls of your feet with your feet close together, keep your hands on your knees and lower your head. Get as low as possible without touching your hands or knees to the ground. DO NOT LIE DOWN!"

I know what you're thinking… "Really, you think I'm going to do that? I'd look stupid!" I agree. But when you watch those TV show about lightning you sometimes see the potential between the ground and air change, almost as if the ground is becoming more receptive to the lightning strike, so assuming you're standing where lightning is about to strike, it kind of makes sense that your hair may stand on end. I've decided if it happens to me, I'm crouching and looking goofy until my hair settles down.

We had a lightning strike near our house one summer. It blew a swath of bark off a huge white pine tree as it traveled to the ground. Then it parted the long grass like a comb through a head of hair as it traveled across the yard to a set of metal steps. I used to run around outside setting up rain barrels during thunderstorms, but after I saw the power of lightning that day, I am now very careful to be safely inside during a thunderstorm.

High-level dangers

These are the types of dangers that pose a major risk to you and your family so you need to take them very seriously. While we don't live in a high-risk area for earthquakes we did experience a mild one several years ago, and it was quite a disorienting experience. It felt like the whole house was swaying and shaking. After watching our bookshelves sway back and forth, I finally got off my butt and secured them to the walls. Michelle had been suggesting I do this for years but it took an earthquake to get me moving. There are a number of ways to prepare for an earthquake and FEMA has lots of good information. www.ready.gov/earthquakes

Be sure to store heavy objects on lower shelves and don't hang pictures or mirrors over beds or where people sit. It's a good idea to secure not just bookshelves but your water heater, refrigerator and even furnace by strapping them to a wall stud and bolting them to the floor. If I had a natural gas line coming in to my home I would check with the gas company to see if they would install an automatic gas shut-off valve that's triggered by strong vibrations. After a strong earthquake, you could still run the risk of a fire and/or an explosion caused by a gas leak, so you should take precautions against this.

Part of your emergency plan, that I'll outline in the next chapter, is to have safe spots in each room under sturdy tables or against an inside wall. I would do more than just write down a plan; I'd hold earthquake drills with the whole family so they knew the best places to go when an earthquake occurs.

Drop, Cover and Hold On ... Drop to the ground; take Cover by getting under a table or piece of furniture, and Hold On until the shaking stops. Make sure you stay away from windows, outside doors and outside walls. Don't use a doorway unless you know it's strongly supported because many are lightly constructed and don't offer the kind of protection you may get under furniture.

While you'll be tempted to try and get out of the building (and I don't blame you since that was my first instinct during our mild earthquake) you should actually find a safe place quickly and ride it out there. Research shows that most injuries happen when people are trying to move too far inside a building or to get out of the building. And this makes sense, because you're trying to walk on shifting ground. You'll just end up looking like those drunks in the movies. And even if they make falling down look funny in the movies, it's another story when it's you. But be logical here. If you're in your front hall when you feel an earthquake, it probably makes more sense to head out the front door rather than trying to get back into the kitchen so that you can take shelter under the table. The odds are though that you spend more time in bed and in front of the TV while you're in your house, so these are the rooms in particular where you should have a spot that you can get to quickly.

Tsunamis

Tsunamis were one of those things most of us didn't have much experience with until Sunday December 26, 2004 when one devastated many costal areas around the Indian Ocean. It is estimated that more than 230,000 people died. The number is staggering so you need to take these seriously if you live near the coast. The 2011 tsunami that destroyed whole cities in Japan was another reminder of nature's awesome power. There was a fair amount of dramatic video footage of the 2004 event but since people in Japan have lots of video cameras and cell phones, the video footage from that event was even more frightening. If I'm ever visiting the west coast and I feel an earthquake, I am getting to higher ground immediately. I'll come down hours later when I'm sure the danger has passed.

The tsunami that struck Japan in 2011 after a 9.0 earthquake off the coast caused 16,000 deaths and up to $30 billion in damages. It also completely disrupted the Japanese power grid after the meltdown at the Fukushima Daiichi Nuclear Power Plant that continues today. If you've seen the video footage of those walls of water rolling through cities you know the potential for destruction of a tsunami. If you hear the sirens, take them seriously.

Tsunamis start with earthquakes in the ocean so a good starting point is to be aware of even the slightest earthquake that you might experience. Even if it seems minor to you at your location it may just mean that the epicenter is a good distance away. Which might mean that it may take longer for a devastating wall of water to appear.

You should find a reliable source of information in your area to get data on potential tsunamis'. I'd start with the website for the National Oceanic and Atmospheric Administration (NOAA) www.noaa.gov. They have lots of great information and you can find the up-to-date reports that you need. They link to services that even provide updates and warning to your cellphone or computer. You can also tune to a local radio or TV station for news.

Ultimately though, you need to trust your instincts on these warnings. In October of 2012 at 8:04 p.m. a 7.7 magnitude earthquake struck off the coast of British Columbia. At 8:13 pm U.S. officials sent out a tsunami bulletin, but emergency officials in B.C. did not send out a warning until 8:55 p.m. That 39 minute delay could have been critical for someone trying to evacuate especially since it was dark which might have slowed people's response times. Luckily no major tsunamis made landfall.

There are numerous layers of bureaucracy, policies and procedures that government organizations have to follow and sometimes their fear of issuing a false alarm can inhibit their ability to get information out quickly.

So if you live on a coast and you feel an earthquake, and think that a tsunami is possible you need to start heeding that funky Stevie Wonder song, "....gonna keep on tryin' 'til I reach the highest ground."

And by all means, don't be one of those people who decides to follow the water out as it get sucks into the ocean. It's gonna come back and it's gonna come back really big and mean. Skee-daddle away from the beach to higher ground as fast as you can!

Hurricanes

I can't think of better proof of Mother Nature's fury than a hurricane. They have always been dangerous and destructive. What we're seeing today though is a new level of intensity. They are made more powerful by more moisture in the air and warmer oceans while they are forming. Added heat and moisture fuel them and this does not appear to be a trend that will end soon.

Hurricanes are one of those things that you can't make many provisions for. If you are in the path of a hurricane you should get out of its way. Unless you have built yourself a reinforced concrete bunker on a hill away from any potential flooding, hurricanes pose a danger to you and you should evacuate.

As someone who doesn't like to do what he's told, I also think this may be the time you listen to authority and do what they say. In fact, I'd even suggest being one of the first ones out. The images of the nightmarish traffic chaos that results when the evacuation order is finally given should be reason enough to try and anticipate it and "get out of Dodge" in advance of the rush to the exits.

Storm surge

A new element of the destructive power of hurricanes is storm surge. The oceans are getting higher for two reasons. As water warms its volume increases so a warming climate is increasing the volume of water in oceans. This warming is also leading to melting of arctic ice, which increases this problem. As storms like hurricanes gather force their high winds can push great volumes of water on to shore. This storm surge was a great contributor to the destructive forces of Hurricane Sandy.

Storm surges can occur at anytime of the year and can even happen on inland waters like the Great Lakes. So be aware that storms and high winds can push great volumes of water on shore anywhere. While it won't have the initial intensity of a tsunami, it has the potential to do a lot of damagc and could put you in danger if you stay in low-lying costal areas or around large bodies of water during intense storms.

Floods

While a flood is usually a gradual process you should still bear in mind they have the potential to cause real danger. It is something you have to be aware of and have an evacuation plan in mind for. During periods of drought it's hard for many of us to imagine a flood, but weather extremes are the new norm, and soil that has been too dry for too long can have difficulty absorbing even small amounts of rainwater, which can increase the risk.

You should have an idea of the flood risk around your home. Authorities often refer to the "100 year flood" as an indicator of whether or not your location is in danger. In other words they're saying it's highly unlikely and if it did ever occur, well, it would so rare it would only occur once every 100 years. The problem is, today's erratic weather makes the 100-year flood just as likely to happen this year as any. Hundred-year floods are happening regularly now. It's easy to be complacent if you think the odds of being flooded are very remote. So if you are in a flood plain, assume the 100-year flood could happen at any time.

This is also one of those times when it makes sense to follow evacuation orders. As rivers start to rise roads and escape routes can become impassible very quickly. And we've all seen those images of people being rescued from rooftops by helicopters. So don't assume the helicopters are coming and make plans to move before it comes to this. I would even go so far as to suggest that if you live in a flood prone area you get yourself a boat. It doesn't have to be a luxury cabin cruiser, just a good strong rowboat. One that floats. And has oars so you can paddle it somewhere. Aluminum rowboats can last forever so even if you don't need it for 100 years, it'll be there when you do.

Tornados

I can remember being terrified as kid of the windstorm in the movie, "The Wizard of Oz." Come on Dorothy, pedal that darn bike! Get home fast! Get in to that storm shelter! It is becoming more and more apparent that taking heed of a tornado warning is a good idea. They just seem to be getting bigger and more powerful and happening more often. Winds can reach 300 miles per hour and the path of destruction can be a mile wide and 50 miles long. In 2011 an EF5 tornado struck Joplin Missouri killing 158 people, injuring a thousand and causing over $2 billion in damage. While watching the news coverage it occurred to me that the damage looked like an atomic bomb had gone off. If you ever needed an incentive to take precautions in advance of a weather event, Joplin should be that reminder.

While certain areas are more prone to tornados they are a possibility literally anywhere in North America. Be aware of the warning signs and weather conditions that precede them. Severe thunderstorms with lots of thunder and lighting and an extremely dark sky, sometimes highlighted by yellow or green clouds are usually what precedes a tornado. If you hear a rumbling or whistling sound it's time to take cover. If you can see a funnel cloud you REALLY need to take cover.

If you are in a house you should go to the basement or take shelter in a small interior ground floor room such as a bathroom, closet or hallway. It seems to me that an awful lot of people who are interviewed after a tornado has destroyed their house took shelter in their bathtub. If you can't get downstairs when a tornado hits find a heavy table or desk and make sure you stay away from windows, outside walls and doors.

More than half of deaths from tornadoes involve people in mobile homes and you can understand why. Mobile homes often aren't well anchored and so do not stand up well to the pummeling of such high winds. If you are in a mobile home or trailer get out and get to the lowest floor of a nearby sturdy building. If you can get to a vehicle be sure to fasten your seatbelt and attempt to drive to the nearest sturdy shelter. If you start getting pelted by debris, pull over and park and stay in your car with your seatbelt on. You should put your head down below the windows and cover your head with your hands or a coat or blanket. If you're driving don't get under an overpass or bridge. You are actually safer in a low and flat area.

Flying debris from tornadoes is what causes most fatalities and injuries so do your best to be in a place where you are out of the path of flying debris. And this may be the time to grab that old football helmet from high school and put it on for old times sake!

Take tornadoes very seriously. Radio and TV stations will be a good source of information. I work on a computer most of the day and I keep The Weather Channel website open on my desktop. When severe weather is approaching the website will display "Watches" and "Warnings" for my area. While they can never predict with great accuracy where and when a tornado might touch down, they can certainly provide a warning that one is possible, and you need to take some time to watch the clouds and horizon and take appropriate action when necessary.

Building a storm shelter or "safe room"

I was watching the news recently after a tornado had done major damage to a community. The news crew asked a family to re-enact the event, and the news report showed them exiting their below-grade storm shelter to find their home completely destroyed. I was in awe of the safety this shelter had provided to them. I'll bet when they were building it they kept asking themselves, "Now just why are we doing this?" And I'll bet the neighbors came over and kidded them about preparing for the coming zombie apocalypse. And then I thought of them huddled in there while a maelstrom raged outside and decimated their house. It appeared they had made a very wise decision.

FEMA now refers to storm shelters as "safe rooms" because they don't want people to just think that you need an underground shelter in your backyard. Sometimes this isn't feasible, but you can certainly have a "safe room" in your house. If you were building a new home this would be easy and fairly inexpensive to include in your plans, but you can also retrofit an existing house.

There are actually full plans for building a safe room on the FEMA website. www.fema.gov/safe-room-resources If you have anxiety about being in a house when a tornado roars through, this may be just the way to reduce that stress. They have a variety of resources on their site to get you started. It is beyond the scope of this book to provide many details but I do think having a safe room or underground shelter at your home makes more and more sense. Make sure you research it well though. While digging a hole and putting in a small concrete block shelter may sound easy, and it certainly isn't rocket science, there are things to consider. You want to make sure it doesn't leak or crack in cold weather, and if it's not anchored properly it can actually have a tendency to want to jack or float up as the water table rises. Now if it floated perfectly square it wouldn't be too bad but you can be sure that it will lean to one side and be a constant reminder that it wasn't done properly.

We live in a farmhouse built in 1888. The basement has foot wide concrete walls, which I think would provide good protection in a tornado. Our challenge is that we live in a very sparsely populated area with our nearest neighbors 3 miles to the east and 6 miles to the west. So I finally put a pry bar, a hatchet and some water bottles in the middle of the basement where we would take shelter. I'm paranoid about being trapped in there if the house collapses, so I put some tools there to try and reduce this source of anxiety.

The eye of the storm

You can never eliminate all risk in life. You could walk out of your apartment building and a piano could fall on you. Or a meteor. During your lifetime though the odds are much lower that a piano will fall on you and much higher that you will be involved in a major weather event. Each year as the planet warms the likelihood of such an event impacting you rises. So you have a choice. Continue to keep hoping you'll dodge a bullet and not be impacted, or start making some basic plans to deal with one when it happens.

Most of the plans don't have to be elaborate. You should just sit down and draw up a plan of action and pick some prime places in your home that will offer the most protection in an earthquake or tornado. You may also want to think about building a safe room or shelter to really increase your odds of getting through a storm. Just get started on the small steps and planning now.

6 Riding Out the Storm – Your Home Emergency Plans

So that approaching rain you saw on the news last night got all fired up with this afternoon's heat and has turned into some pretty nasty weather. As its force increased over time it started taking out power lines and phone lines. Tornados are touching down. You're at work. The kids are at school and your elderly parents across town are not going to be reacting well to the loss of power for your mom's oxygen machine. So what's your plan? You've got a plan, right?

Most of us don't have a plan because we aren't put in these situations very often and this is a normal human reaction. We are conditioned to deal with imminent threats. If we don't have to deal with these scenarios very often we don't instinctively have a mechanism to handle them.

So you need a plan, and as hokey as this sounds, you should write it down. I know, I know, that's what nerds do and you are way too cool to be putting on paper a plan to deal with an emergency. If you go down that road what will you do next? Start cataloguing your shoes to match your suits?

As awkward as this sounds it's a good idea. By having it in writing you'll have a plan that you've thought out when you're calm and cool. Unless you're a cop or firefighter who spends every day dealing with a crisis, you may not react well to an emergency. If you've written it down even if you don't refer to it, at least you've thought through a plan of action. This is half the battle. You've got your brain working in the right direction.

A great place to start is the FEMA website. (www.fema.gov) They have a "Make a Plan" page with a great "Family Emergency Plan (FEP)" form you can download. And yes, I know, you're busy, who has the time to sit down and fill out two more pages of information like another credit application? You do. You just have to make it a priority. This is way more important than that credit app.

On the form you record where you plan to meet, as well as the name, date of birth, social insurance/social security number and important medical information for each family member. You include the addresses and phone numbers of where everyone spends their day... work, school, etc. There is a second page where you record everyone's phone number. Be sure that each family member has a copy of this information in his or her possession.

Communication

Remember that phones often don't work during emergencies. Many of us no longer have landlines which are lines which come from utility poles into our homes. We just use cell phones which are wireless and get their signals from cell towers throughout the landscape. Cell towers are well made but often even if cell phones are working it's really tough to get a line because guess what? Everyone else is in the same boat and is trying to call someone. Cellular networks create capacity based on how many calls are expected at any given moment on a regular day. The hours after a tornado blows through are not normal, especially with phone volume.

Remember that sometimes even if you can't get through to someone in your community you may be able to get an outside line to someone in another part of the state or country. So your plan should have a dedicated person that family members can call outside of your community to try and coordinate information. I know it sounds crazy, but if you have a family member or friend who is pretty consistent at answering their phone, then ask them if you can designate them as your family's "Go To" person and make sure your written plan has their number front and center.

We finally made a "Family Phone Number" page that we update regularly. We have one in the car, one on the wall near the phone, one in the emergency kit, in Michelle's purse, my briefcase ... everywhere! We bought some bright pink fluorescent paper to print the phone list on. This way it's easy to find. We started making this list because our daughters kept changing their cell phones when they were at university, then moved to other cities for new jobs. And with the competition in the cell phone business it's easy to change carriers regularly. So make sure that plan has everyone's 'current' cell and telephone numbers. Some people change phone plans like socks, so make sure you know what number they currently use.

The FEMA Website has a "Child's Emergency Contacts Card" which you can download and fill out by hand if you don't feel like making up your own.

Text Messaging

Can you text message? If you're like me and over 50, the answer is probably no. I am "Text Message Challenged." For years I have been conducting a *"Thriving During Challenging Times"* workshop at colleges and I always recommend that people know how to text message. This is because text messages use a less congested part of the cellular network. It is more resilient than the bands they allocate for voice communications, and as mentioned those voice lines will probably get jammed up very quickly as every one in town (and out of town) starts calling during a crisis. So even if you can't talk to someone on their cell phone you may be able to get a text message through.

My confession is that for years I told people to do this, but I didn't actually know how to do it myself. So when we got our most recent cell phone I forced myself to learn. I've worked in computers since 1984 but I had a mental block about text messaging. I had to work through menus and every time I went to send a text my brain went blank and I couldn't remember the process. If I had a dollar for everytime I typed a text message then lost it, well, I'd be writing this book from my yacht. So get a cell phone and learn how to text message and practice. Every time you leave the house for a month text message your kids. Send them a text once a month just so you remember how. It's cheap, they'll love to hear you tell them how great they are, and you need the practice until it finally imprints on your brain.

Institutional Plans to be Aware of

If your kids are at school or some other organized event, the institution will probably have an emergency plan of its own. So put on your 'nerd' hat and go to the office and ask what the plan is. If they plan to keep your children at the school until a parent picks them up, even if it's past the regular bus time, then write that down on your list. What if the windstorm blows a big tree down on top of your car and you aren't able to drive? Designate someone as a backup who will pick your child(ren) up in an emergency. Perhaps you could ask the parents of one of your child's classmates. And you could agree to do the same for them. Both the school and your child must know who this person is.

The thing I like about the FEMA "Family Emergency Plan" page is that it suggests three locations:

1) Neighborhood Meeting Place
2) Out-of-Neighborhood Meeting Place
3) Out-of-town Meeting Place.

Your family will logically return to your home if they can, but what if it's flooded? You need to pick another place nearby in your city to meet up. Maybe city hall or the library, especially if it's on higher ground. And remember that some natural disasters we're experiencing end up being much more widespread than anticipated. If you were in New Orleans during Katrina both your Neighborhood Meeting Place and Out-of-Neighborhood Meeting Place may have been out of bounds. So having an Out-of-Town Meeting Place would have made perfect sense. Pick Grandma's house in the next county over. It may take every one a few days to get there, but at least you'll know where everyone ultimately plans to end up. This will hopefully reduce some of the stress at a really stressful time.

Once you've got this plan written down, remember to discuss it with the kids. Not in a screaming "IT'S THE END OF THE WORLD AS WE KNOW IT" tone, but in a "in case we ever have one of those big wind storms blow through, here's our plan to get you." Be mellow and they won't be all freaked out about it. My experience is that my kids are always more "yea whatever" about this stuff than their parents. FEMA has a section on their website called "Ready Kids" and if you have younger children you may want to try out some of the activities just to ease into this.

Once you've successfully managed to get every one home you should itemize the things you have to do. In a black out it's recommended that you turn off all major appliances. Often as the utility company works to get the electricity grid back on line there can be spikes and inconsistent power. You can't really blame the utility because our electrical grid is one of the most complex machines humans have ever built. By turning off as many loads as you can you actually make it easier for them to get back up and running. Once the grid has been stable for a while you can start plugging things back in.

Your plan should also indicate where the major shut offs are for other utilities that come into your house like natural gas and water and where your electrical utility box is. You should know where the floor drain in the basement is. This is where water would flow out of your basement in a flood. If there's a carpet in the way it may not work as well it's supposed to.

By listening to the radio you may get guidance from the local authorities on whether you need to deal with any of these other utilities.

Natural gas should be something you're aware of especially if you're in an area prone to earthquakes or tornados. Since natural gas lines are underground they could conceivably rupture during a natural disaster and since natural gas is explosive, it's not something you want leaking into your home. If you smell natural gas in your home, get out quickly. If your home is an area where there has been an earthquake you may want to shut off the natural gas coming into your house.

For good instructions on how to shut off your natural gas be sure to visit this website for the most current information;

www.ready.gov/utility-shut-safety

Remember, never turn the gas back on yourself. Get a qualified technician from your natural gas supplier to do it for you. The reason is simple, they will have the tools to do it correctly and ensure that there are no leaks that could be hazardous to you and your family.

The best emergency tool you can have!

Everyone in the family should be aware of where you keep your flashlights. You should have a number of really good ones and it's important to make sure that they work! It's worth investing in some of the good quality "LED" type flashlights. The old ones used bulbs that required a lot of energy so the batteries never seemed to last very long. LEDs require a fraction of the energy, which means they'll last much longer. We also invested in a rechargeable 1million candlepower hand held spotlight which is incredibly bright. It has a lead-acid battery so we charge it up every couple of months to keep it in good working order.

I believe one of the smartest tools you can have for an emergency is a "headlamp." This is basically a flashlight that you strap on your head, like what you might see miners wearing. I would have to say this is probably one of my greatest tools for living off the electricity grid. While my system is working very well now, it always seemed wasteful to me to light up large areas outside if I only really needed to see where I was working. If I go out the woodshed at night to get firewood I could turn on lots of floodlights and illuminate the path to the woodshed, but I'm only actually on the path for a few seconds. Most of my time is spent in the woodshed. So by having a headlamp I just light up where I'm working or walking and save all that wasted energy. In a power outage you won't have that option. Even though I do have the option, I still opt for a more efficient headlamp, so you should follow this example.

You can spend a lot of money on good quality headlamps but I got one of my favorites at a dollar store and it's worked for years. As long as I keep the batteries charged, it's just an awesome tool! It's so much better than a flashlight because it allows you to have both hands available for work, as opposed to having to use one hand to hold a flashlight. In fact I'd go easy on regular flashlights and have a strap on headlamp for everyone in the family. Sure, you'll look like geeks all walking around the house with headlamps, but come on, it's a blackout, who cares?

Fire extinguishers are an important safety tool that should be in every home. Family members need to know where they are located and how to use them. Again, showing a child how to use a fire extinguisher can make a lot of parents nervous like it's just one more thing for the child to worry about. On the other hand that child may have seen fires on TV or know of one in the community so be aware of them anyway. Just like strapping on a seat belt in a car, you're merely showing them a tool that hopefully they'll never need, but which will come in handy if they do.

Here's a great video to watch with your child so they can see how to properly use a fire extinguisher.(www.youtube.com/watch?v=Hw4uIiXUCY4)

If you have a fire and there are two of you at home only one person should try and extinguish the fire. The other person should call "9-1-1". Then use the P.A.S.S. system.

Pull the plug
Aim at the base of the fire
Squeeze the trigger
Sweep the fire

If the fire is past the very beginning stages or is too big and past the realistic point of you being able to extinguish it, get out!

Pets

Your plan should include how to deal with your pets. If you have to evacuate and head to a shelter there's a good chance they won't take pets, so have a kennel or other location where you can take them.

Always keep a cat or dog carrier handy so that if you have to evacuate you can easily load your animals into their carriers and this will help to keep them safe and confined during a stressful situation. Remember to include pet food and water for your pet in your plans.

Home Emergency Kit

The most important part of your home emergency plan will be your Home Emergency Kit. Your kit may consist of stuff you already have in your house but what you need to do is gather it all up and store it in one spot. Remember that emergencies often occur when it is dark and wet and cold and disorienting. Having everything together in one spot will just make it easier if you ever need it.

You can purchase emergency kits in many stores or from organizations like the Red Cross (www.redcross.org.) You can also put your own kit together. It should be cheaper and will just have items that are most relevant to your family.

I'm going to address storing larger volumes of water in Section Three, but you must remember that each person in your household will require 1 gallon (3.8 L) of water to drink per day. If you've got a family that can be a pretty big volume to begin with, especially if you're planning on having enough for 3 days without regular city services like drinking water. So you'll need 12 gallons of water to start for a family of 4. And this is just to drink. If you hope to wash and clean your teeth you'll need more.

For now plan on having a number of cases of bottled water, or the larger 1 gallon (3.8 L) or 2 gallon (7.57 L) jugs, enough for a gallon per person per day. If you're frugal like me watch for these types of containers in your neighbors' recycling bins. They are plentiful out there and there's no reason not to grab clean ones, rinse them out and reuse them. If you fill them with tap water you have the added advantage of a bit of chlorine having been added by your municipality to kill pathogens which may help if the water sits for a long time before you use it. Store your water near your emergency kit if you can.

You might want to get a big clear plastic storage container with a lid to put the bulk of your emergency kit in. If family members can see it every time they walk by it, they're more likely to remember where it is if needed quickly.

Here is a summary of what your kit should include. I'll explain some of the items below;

- Portable Radio with spare batteries
- Waterproof Matches and lighters
- Candles
- First Aid Kit
- First Aid Manual (know how to use it)
- Food (canned or individually packaged & precooked)
- Manual can opener
- Infant Supplies (if applicable)
- Critical medication and eyeglasses
- Flashlight/headlamp (spare batteries)
- Dust Masks
- Spare Cash – including coins for payphones if they still exist in your area
- Duct Tape
- Basic Tool Kit
- Blankets or sleeping bags for each family member

Cooking Supplies

- Grill – only to use outside
- Small Pots & Pans w/utensils
- plastic bags (various sizes for trash and waste disposal)
- Chlorine bleach and powdered chlorinated lime (add to sewage to disinfect and keep away insects)
- Toilet paper
- Toiletries (personal hygiene supplies, toothpaste, feminine supplies, soap, etc.)
- Infant supplies for cooking (if applicable)
- Newspapers (to wrap waste and garbage)

Portable Radio

Having a portable radio really important. It will allow you to monitor what's happening, get a handle on how bad things are, know how large an area the disaster has effected and how soon you can expect help to have the power restored. A wind-up radio is perfect because it will run

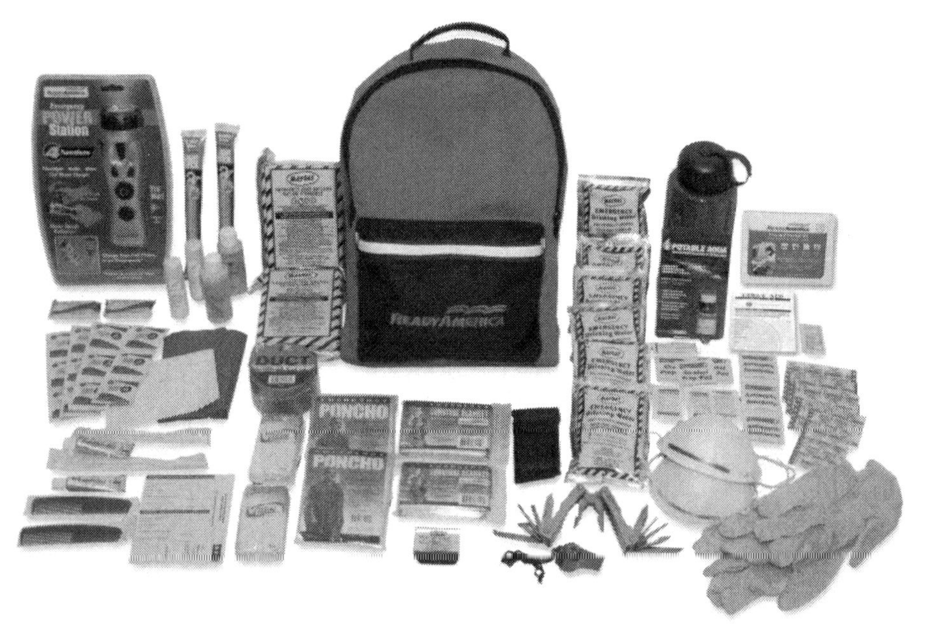

Here is a 2 Person Deluxe Emergency Kit (3 Day Backpack)
www.readyamerica.com

on batteries or you can wind it up if the batteries lose their charge. You can get solar powered ones too but it's hard to remember to keep them charged and if you're in the middle of cloudy weather or you have smoke from brush fires, you may not get much solar power.

Matches and Candles

Light is a great thing and candlelight can actually be pretty calming, so have some candles and matches or lighters in the kit. Remember to be extremely careful with candles. Put them in aluminum pie or muffin tins so if you forget to blow them out and they burn right down they won't set your house on fire.

First Aid Kits

You can always buy First Aid Kits and spend lots of money, or you can make your own. Some of the items in a First Aid Kit are not necessarily that common, so this may be a case where it's worth investing in a premade kit.

Here's what the Red Cross suggests you have this in your First Aid Kit:

- Emergency telephone numbers for EMS/9-1-1, your local poison control centre, and your personal doctors
- Sterile gauze pads (dressings) in small and large squares to place over wounds
- Adhesive tape
- Roller and triangular bandages to hold dressings in place or to make an arm sling
- Adhesive bandages in assorted sizes
- Scissors
- Tweezers
- Safety pins
- Instant ice packs
- Disposable non-latex gloves, such as surgical or examination gloves
- Flashlight, with extra batteries in a separate bag
- Antiseptic wipes or soap
- Pencil and pad
- Emergency blanket
- Eye patches
- Thermometer
- Barrier devices, such as a pocket mask or face shield

I love shopping at thrift stores and I have found a bunch of great first aid books. I hope I never have to stop a major bleeding wound or set a broken limb, but the reality of a disaster is that not only are the odds higher that you'll come across someone who requires this, but it's even less likely professional help will be readily available. At least having a first aid book will give you some basics. I would also suggest you take a first aid course. They are often offered through groups like St. Johns Ambulance and are a great investment of a Saturday morning or Tuesday night.

Food

I think after you finish reading this book I'll have convinced you to have a well-stocked pantry. Part of that will include canned goods, so having an extra can opener in your kit is an excellent idea.

Infants

If you have infants in your home you'll need to scope out what you'll need for three days in terms of formula, diapers, etc. We used cloth diapers on our daughters and I was a big fan, but we also kept disposable ones for car trips. This is one time when having a reasonable supply of disposable diapers makes sense, because you don't how long it will be until you can conveniently wash cloth ones. If you use cloth diapers, having some disposable diapers in this case would be considered a "redundancy" which is a key to good emergency planning.

Misc.

If someone in the family needs essential medical supplies, you should store some in this kit. And if you have an old pair of eyeglasses that aren't too different from your current prescription, include them in your emergency kit in case you lose or break your best pair. Dust masks are handy if you are in an area prone to earthquakes or wildfires where there may be a lot of particulate in the air. I think keeping some cash in your emergency kit is a great idea. When the power is out all those electronic payment systems DON'T work, so you'll need cash. Having some coins for payphones if you venture out are a good idea too. In my small town there is only payphone and it doesn't take coins anymore, just phone cards, so you might want to scope this out beforehand. Do you know where the nearest payphone is?

Having an emergency kit like this is good discipline for forced savings! And always have duct tape in the kit. You can fix just about anything with duct tape!

Sleeping Bags

If you're in a "shelter-in-place" situation you will probably have blankets available. Make sure that you have enough of them and that they are stored on your upper floors, not your basement, especially if it's prone to flooding. You can buy sleeping bags for very little money these days. I actually took a cheap bulky one on a canoe trip very early in the spring one year and was toasty warm while my buddy who had spent several hundred dollars on an ultralight, high-tech sleeping bag froze his butt off. I would suggest next time you see them on super sale pick up a few and store them on an upper floor of your house. If a twister blows through and a family from the down the street is bunking at your house think of how comforting it will be for them to curl up in sleeping bags, even they are on a floor.

Cooking

Preparing food can be challenging in an emergency. The nice thing about canned goods is that in a worst-case scenario you can eat the contents without heating them up. In fact I prefer baked beans right out the can and not heated. Depending on the emergency, even if the electricity is off your natural gas company may continue to keep your home supplied with gas for cooking. Your cookstove will likely have an electronic ignition that causes a spark to ignite the gas flame. This won't work in a blackout so you should have matches or a lighter available to start the stove.

If you cook with electricity or don't trust that you'll have gas then you should think about other ways to prepare food. If it wasn't a tornado that caused your power failure, and took your back porch with it, then you can always use your bar-b-que. If it's winter you can still use it but only to COOK OUTDOORS. If you use it indoors not only could it burn your house down but it could fill your house with lethal carbon monoxide gas. This is why it can make sense to have a backup propane tank for your bar-b-que, just to ensure you can cook outdoors during an emergency.

If you are a camper you may have a camp stove. These are great but remember again to use it outdoors. Having a few cans of kerosene or campstove fuel, or the small propane tanks if that's what the stove uses, is recommended. You can also purchase small stoves that run on hard fuel like "Sterno" that are quite economical. Stock up on "Sterno" and remember to test the stove a few times just to make sure you're comfortable with it and have realistic expectations on how long it takes to cook things. Ensure you have pots that are a suitable size for these stoves.

That Whole "Sanitation" Thing

One thing that advanced society has done has removed us from having to deal with the reality of the waste we produce as humans. So during an emergency this becomes a really big deal. Human feces can harbor viruses, bacteria, parasites and all sorts of nasty things you don't want people coming in to contact with. During a stressful situation where people may not have proper food and water and sleep, their immune systems will be suppressed which will make them all the more vulnerable to these pathogens which might make them sick. So you need a strategy to deal with it. Believe me if you ignore this section, two days into a crisis your bathroom will not be a hospitable place. The news seems to be full of stories about cruise ships that become disabled. Dealing with human waste quickly becomes a huge problem. No power to pump water meants no flushing toilets for thousands of people for days. They are not happy campers!

First let's consider a short-term, shelter-in-place scenario where you are able to stay in your home. If the water lines are broken but the sewer lines seem to be working you can flush the toilet by pouring a bucket of water down it. This only works if you have a large supply of water available like a pond or lake, swimming pool, hot tub or the ocean. If you have lots of rain barrels available you can use them too, but remember that ultimately if the water stays off for a long time, you might want to save some of this water for cooking or washing. So only use the bucket method if you know that water won't be a problem.

Now let's assume the sewer and water lines are broken so no water is being provided to your home and you can't flush the toilet. Well, you will be able to flush it once, but once the water that's in the back of the tank is used up, you have a problem.

Your emergency kit should include a large number of plastic bags that fit in the inside of your toilet bowl. Then you can use the toilet as you normally would. When you're done add a small amount of disinfectant to the bag and seal it. You can use twist ties for this. If it looks like your emergency should be short-term plan on just using a bag once. You should put the bag in a covered container like a garbage pail in the garage or outside.

A Luggable Loo from www.relianceproducts.com

What if a tree blew into your bathroom and broke your toilet? Well, while it doesn't sound like fun you can use the same plastic bag method with a bucket. I suggest you buy a sturdy bucket that you keep with your emergency kit and store some of the plastic bag liners inside. You might want to put the bucket in something like a milk crate to keep it from tipping over because let's face it, balancing over a bucket would be a challenge for a gymnast. For the rest of us it's just a recipe for disaster. Or how about using that foam pipe insulation you bought but never got around to putting on your hot water pipes. Put it around the rim of the bucket to make it more comfortable. There's nothing like a good 'ole crisis to bring out people's ingenuity!

An even better suggestion would be an actual portable toilet like the one you see in this picture. (www.relianceproducts.com)

Load it up with plastic bags, store it near your emergency kit and hope you never need it. Reliance provides "bio-blue" with their "luggable-loos" which is a dry powder deodorant in pre-measured pouches.

You can also make your own. The easiest disinfectant to use is liquid chlorine bleach, which you can use to make a solution of 1 part bleach to 10 parts water. You could use sawdust from the workshop or even kitty litter. Ashes from your fireplace might work but you have to be really, really careful with ashes. Even if you haven't had a fire for days there may be glowing embers in them, and if they come in contact with plastic they could quite easily start a fire. And this is not the time you want to be dealing with that. Some hardware stores sell powdered chlorinated lime to disinfect and deodorize.

Latrines

If you are in the country you may want to consider digging a latrine. While these are usually frowned upon by government authorities, for short term use during an emergency where other arrangements can't be made it's an option. If you have a well for drinking water keep the latrine as far away from that as you can. Dig it 2 to 3 feet deep and pile the soil from the hole in a pile right next to it. This will allow you to backfill a few handfuls everytime it's used. This is where a well-stocked garage or garden shed with hand tools is so helpful. A good shovel will come in very handy in an emergency.

You must consider the disposal of human waste seriously. You'll have enough problems to deal with and you don't want to add a preventable illness to your list of concerns.

Toilet paper

If you are one of those people who only buys toilet paper when you run out of it, it's time to break yourself of this habit. Start stocking up on it, NOW. Next time you see it on sale, put how much you think you'll need for the next month in your cart, and then double or triple the amount. It lasts forever and it's just too great a luxury to be without. When the power goes out and stores close, this is the kind of stuff that quickly sells out. Since so many people now get much of their information online I can't even suggest newspapers and catalogs as a backup anymore. And they are an extremely poor replacement even if you have them. So stock up on toilet paper.

I had been stocking up on an inexpensive toilet paper I like, when Michelle brought a new type home recently. Well it wasn't new, it was the same product it just had twice as many sheets on each roll. So the price was better. What I liked the most was that you ended up with twice the amount of toilet paper, but it takes up almost the same amount of space. It's a much more efficient way to purchase and store toilet paper. This becomes especially important if you have limited storage space or live in an apartment. So always be experimenting and looking for better options for supplies like these.

Toiletries

I'm not a big fan of hand sanitizers but I think they'd be a great thing to have in your emergency kit. Of course have some extra soap, but remember you need water for soap and it might not be readily available. Your hands are going to get grungy and hand sanitizers will be a welcome bit of civility at a bad time. And have a supply of feminine hygiene products and little luxuries like toothpaste. Again, as you're building your emergency kit, look for these items on sale and buy 2 or 3 times what you'd usually buy and throw the extra in the kit. These products last a really, really long time so there is no downside to this. You can always use them eventually.

7 What's in "your" bug out bag?

Last chapter I talked about a "shelter-in-place" emergency, where you ride out the storm and resulting crisis in your home. But what if that's not an option? What if you never did get around to following up on the suggestions in the "Where to Live" chapter in Section III and now a hurricane is bearing down on your costal home? Or what if even though you'd scoped out the perfect place to live, a brutal drought has caused a wild fire that is heading towards your house?

If a storm is approaching it's important to plug into the media machine and get a handle on what's happening. Often if weather conditions are setting themselves up to be messy later in the day as the temperature rises, weather forecasters will be able to give some warning. They risk the whole "boy who cried wolf" syndrome if they try and get too sensational about the odds of a dangerous weather event, but I think they'd rather err on the side of caution and at least let people know it's possible. With larger weather events like hurricanes we now have days of warning so there's no excuse not to have some idea of a pending evacuation notice.

If you are in an area that might be affected by a hurricane or flood, authorities may ask you to leave. They'd rather you get out while you can on your own, than to have to send in boats and helicopters later. Often news reporters will find a defiant person who vows to stay put. I guess this is their prerogative and it may be yours. I know I would hate to leave my house. You have to consider the nature of what's coming and how well you think you'll be able to ride it out. If you've got a good emergency kit and food and water stored high in your home, then maybe it won't matter if floodwaters start to rise around your house. Or better yet, if you have your own rowboat tied up firmly to the house you know you have an exit strategy.

You should resolve to **not** be that person who needs rescuing. First off, you're asking someone else to put himself or herself in harm's way because you chose to ride out the storm and ignore the warnings to leave. I believe they refer to this as a "moral hazard."

We saw it with the economic collapse of 2008. Banks behaved very badly. But we bailed them out. That's what moral hazard is, because when you bail out the banks when they behave badly you reinforce this behavior. If they think they're too big to fail and the government will always bail them out, then why wouldn't they make risky investments that might increase their bonuses? There's no downside to it.

It's the same with emergency services. If you warn people to leave, and they don't, at what point do first responders consider the risk to themselves, trying to rescue people who chose to stay? From watching disaster scenes on TV I think first responders rarely think this way, but it's always a risk you take. And do you really want to be the one getting hoisted up to a helicopter from your roof when your house is surrounded by water and your escape routes are gone? That whole putting the foam harness under your armpits and holding on for dear life looks kind of terrifying to me. I know I haven't been training for it.

And let's not forget what the new head of FEMA, Craig Fugate, said shortly after taking over the job in 2009; "We tend to look at the public as a liability. But who is going to be the fastest responder when your house falls on your head? Your neighbor."

(Check out the article here; www.theatlantic.com/magazine/archive/2009/09/in-case-of-emergency/7604/)

He seems to be suggesting that FEMA is starting to see the moral hazard effect. If they always arrive within 24 hours with food and water, then people will always expect that. And ultimately in a world with multiple climate change disasters citizens have to start being more resilient. Hence, you reading this book! Thanks and way to go!

So, it looks like it's time to leave.

Emergency Car Kit

You may already have a home emergency kit, but it doesn't hurt to have many of these items in more than one place. This is redundancy. If the tornado takes your house but leaves your car, you have a starting point for your recovery.

Here are some things you should have in your trunk just for day-to-day driving that will help in an emergency.

- jumper cables
- Fire extinguisher
- flares/reflective warning sign
- windshield washer fluid
- a bag of sand or salt (you can use non clumping cat litter)
- small shovel
- snowbrush
- blanket
- first aid kit
- tool kit

So here are some ideas for an additional emergency kit:

- food
- water
- cash
- candles
- extra clothing and shoes
- whistle
- flashlight (wind up)
- extra clothing
- an extra copy of your emergency plan
- multi-tool

Okay, okay I know what you're saying. "How will I have any room for groceries? I did the responsible thing and bought a small car and these two plastic containers are taking up half the trunk." Sorry, I can't be much help here. I've had to use my jumper cables quite often, especially during cold weather. I've never needed my fire extinguisher but I've seen a car on fire and I can't think of a worse feeling than not being able to help someone in a fire. Although I try and keep my windshield washer antifreeze reservoir full all the time, it's Murphy's Law that the day it runs dry I have a long drive on a wet, slushy highway. And if you have to stop and buy a jug of antifreeze at a highway rest stop they charge at least twice what you'd pay at the hardware store. So have a spare unopened one in the trunk. I say unopened because they invariably fall over and tend to leak once you've taken off the airtight seal.

I always have a blanket in the car and it has come in handy for packing breakable stuff so this is a handy thing to have. During winter storm season many people get stranded in their cars, sometimes overnight. And a first aid kit is a great idea. Again if you're helping someone injured at the side of the road it sure would be nice to have some gauze and tape to wrap around that wound temporarily.

With the price of tools these days it only makes sense to have a tool kit in your car. Sometimes you can find a complete kit for a ridiculously low price on sale. I suggest you have one in the car, even if you're not necessarily a handy person. Because even if you can't figure out what tool you need someone who stops to help you might need to use a tool.

Be sure to keep a bag of sand in your car, especially during the winter. Make your own sand bag next time you're at the beach! And a shovel with a collapsible handle will seem like a huge waste of space, until the day you miss that turn and end up in a snow bank. Using the snowbrush and your feet in a blizzard to move snow just does not work as well.

We get a fair amount of snow where I live and so we have a complete set of snow tires on rims that we put on and take off each winter. Having them on rims is easier on the tire and cheaper to put on and take off. I have driven with all-season radial tires in the winter and there is no comparison to snow tires. Snow tires are infinitely better in the snow. If you're in a location where snow is the exception then it may not be a good investment, but if you live in Syracuse where you get those endless streamers of lake effect snow blowing off Lake Ontario, they make complete sense.

More on the car emergency kit

You can use sturdy cardboard boxes with lids to store your emergency car kit items. I invested in two rectangular clear plastic containers with lids. I like them because I can see what's inside without having to open the lid and root around them. I like the ones with squared edges because they are more space efficient than round ones. When you put a couple of round ones together you end up with a lot of waste space, whereas square ones fit tight together with no wasted space. When you have a small car and have as much stuff in the trunk as I do, this can be a big deal.

Food is one of the items on the list for your car emergency kit. We've all heard the stories of people who go off the road in a snowstorm and live for days on granola bars. You might seem kind of eccentric, carrying granola bars around with you, but those stories prove that they work! Choose food items that are energy dense, well packaged and individually wrapped. Trail mix, crackers, nuts, energy bars, granola and dried fruit are all good ideas.

Choose food items that are not too expensive and that you can cycle through periodically. Items stored in your car are going to experience greater temperature extremes than you'll get in the house, so they won't last as well. And if you're buying things like granola bars try to avoid ingredients that will melt, like chocolate and yoghurt coatings. The best test is to put some in the trunk in July and then eat them the following month. If they're a melty mess, they didn't pass the "hot weather" test.

Water will be the opposite in that you'll want to test it in January. If you use water bottles see how they respond to freezing solid and then thawing. After a few freeze/thaw cycles the bottle and water may not look that appetizing, but remember this is for an emergency. In times like those we easily drop our standards of what's acceptable to eat and drink.

Cash seems like a strange thing to have in your car but I think it's a good idea on so many levels. First off, most of us are out of the habit of carrying much cash in our wallets and purses on a good day. On a bad day, when the power's out and things are amiss, those credit and debit cards will not work. Electronic systems can crash even without blackouts. Cash works even without electricity! I am terrible for forgetting my wallet when I go out. So I always find it comforting to know if I get halfway to town and realize I've forgotten my wallet, I've got $50 for food or gas or whatever I need. Storing cash sounds scary, but it's pretty unlikely anyone is going to break into your car to steal some cash (which will be hidden.) And finding out that you forgot your wallet at the cash register is way less embarrassing when you can run out to your car to grab your cash stash. I keep my $50 stash in an envelope in the springs under my car seat.

As I've mentioned, I live out in the country and my home is not hooked up to the electrical grid. When I arrive home on a dark night there are no streetlights and often I don't want to "waste" electricity on outside lights when I am not at home. It can be so dark that we literally cannot see our hands in front of our faces. Getting from the car to the front door is a slow, hazardous walk. To avoid this I keep several good flashlights in the car that are readily accessible. I have some battery-powered ones as well as a wind-up version. I never know when the batteries are going to pack it in on the regular flashlights, so it's nice to know I have one that I can count on working. I love LED flashlights. The bulbs are rugged and so efficient a good set of batteries seems to last for years even with fairly regular use.

Having extra clothing in your car may seem a little like overkill but you may find yourself cold or wet at some point and a change of warm, dry clothes will feel like a luxury. Are heavy rains or floods common in your area? Imagine having your house destroyed by a hurricane and having to stand and look at the damage in cold, wet clothing. Think of how miserable a feeling that would be. Think of how fantastic a change of warm, dry clothes might be. Yes, you may never need them, but like any insurance, you really hope you never need to make that insurance claim, but if or when you finally do, it's pretty great to have.

Having an extra copy of your emergency plan would be a great thing to have in your car's emergency kit. It will contain all the information you thought through when you were calm and had a clear mind. It will also have emergency numbers, friends, school, work numbers, all those things you need most just when you're least able to reach back in your stressed brain to come up with them.

A well-stocked car makes some of the stuff in your bug out bag redundant, and this is a good thing. Redundancy. You're putting together your own personal "Plan B." And having a "Plan C" can only help in the time of a crisis. Best-case scenario it's just extra stuff to use to help your neighbor.

Bug Out Bag - BOB

Now we should talk about Bug Out Bags (BOB). There is no shortage of names for them; "Flee bags," a "Grab Bag," a "Ready to Go" kit, a PERK "Personal Emergency Relocation Kit," "Get Home Bag" or a "Go Bag." There are also some that I would group into a more survivalist mode like a "Get Out Of Dodge" (GOOD) bag.

I think of a Bug Out Bag as something you would grab as a hurricane approached. It would be prepared with the assumption of you being away for 72 hours or 3 days. The assumption is that after this time help, perhaps in the form of organized government agencies like FEMA will have arrived, or else you'll be able to get back to your home. And of course you're hoping your home will still be intact.

A "Get Out of Dodge" kit on the other hand has more the feel that you might have to flee your home and that there is no indication of when or if you'll be able to come back. In other words, you need to have tools to survive in the woods indefinitely. This kit will have more items, be much heavier and assumes you will have a certain level of personal familiarity with survival skills. There's no use taking a small saw to cut up small trees, unless you've built yourself a shelter in the woods before. Sure you can learn, but it would be best if you had tried it a few times in advance of this. Many survivalists refer to needing one of these heavy-duty evacuation kits when the SHTF (sh*t hits the fan). While this is always a possibility

I believe it's less likely than you having to evacuate for a natural disaster. And yes, zombies might one day attack and you should eventually be prepared but for now, let's just assume you'll be gone 72 hours.

I'm going to focus on Bug Out Bags since I think they are what most of us will need and they are a good starting point.

Contents

Here are some basic items you'll want to have in your Bug Out Bag;

- Food
- Water
- Water purification system
- First Aid Kit
- Flashlight
- Head-lamp
- Cash
- Matches and Lighters to start a fire
- toilet paper
- I.D.
- Plastic tarp
- a copy of your emergency plan
- sleeping bag
- emergency blanket
- duct tape/rope
- multipurpose tool
- required medicine
- weather clothing – rain gear
- Gloves
- hand crank radio
- maps of the area
- whistle
- sanitary hand wipes or that alcohol based sanitizer
- N95 or surgical Masks

One of the biggest challenges you're going to have with a "Bug Out Bag" is figuring out how big the bag itself should be. Maybe you have some old backpacks lying around that your kids have lost interest in. If you don't mind 'bugging out' with a glorious pink Sailor Moon backpack, then by all means, use one of the old ones. What you'll probably find though, if the backpack was designed for schoolbooks, you won't be able to pack many of the supplies you'll need in it.

The option at the other end of the price scale will be one of those $300 Arctic expedition backpacks that look like something a spaceman would strap on for a walk into space. Although these are no doubt good quality and well thought out in terms of their design and storage, I don't think it's logical to invest in a pack in that price range. Remember, we're trying to be "sensible" preppers. Unless you have $300 to spend, then sure, go crazy. I would look at an intermediate sized and priced backpack. It can basically have just one large storage area, and will hopefully be bigger than a school pack.

This may be one of those items you add to your "Salvation Army/ Reuse/Thrift Store" list to watch for. We make the rounds of various thrift stores which sell donated stuff and their sporting good section is usually full of school backpacks but once in a while I'll find a half decent larger pack. Garage sales are a good spot too since backpacking is often one of those things people "get into" and then realize it involves hiking and walking and bugs, so a couple of years later they put it on a table in the driveway on a Saturday morning with a $5 price tag.

Be careful of those 1970's- style packs with an aluminum frame. I had one for high school backpacking trips and while the frame looks logical and rugged, the metal always seemed to rub my back in various uncomfortable places, which was a real annoyance. Newer backpacks have removed this 'encumbrance.'

http://www.overboardcanada.ca/pro-sports-waterproof-backpack-30ltr-red.html

If you are going to buy a new bag it doesn't have to be the most expensive or highest quality one available. Something from a discount store is probably fine. Remember you're not hiking across the country. You just want a bag you can grab quickly that will tide you over for 3 days. If some of the stitching isn't of the highest quality, then just reinforce it yourself by hand. Try and find a pack that is waterproof. You'll want to keep all those great supplies dry. And make sure the shoulder straps are padded which will reduce fatigue and will be less likely to cause discomfort.

There are lots of websites that offer backpacks and other similar gear;

www.gofastandlight.com

www.gossamergear.com

www.hyperlitemountaingear.com

On the previous page is a backpack I like.

http://www.overboardcanada.ca/pro-sports-waterproof-backpack-30ltr-red.html

It's waterproof - really waterproof - in case there's a flood coming your way, has comfortable-looking straps and holds 8 or 9 gallons (30 liters) worth of stuff. While you can probably get away with 5 gallons (20 liters) of room inside to store stuff, I'd go for at least 8 gallons.

Here's a photo of an "American Red Cross Personal Disaster Survival Bug Out Kit."

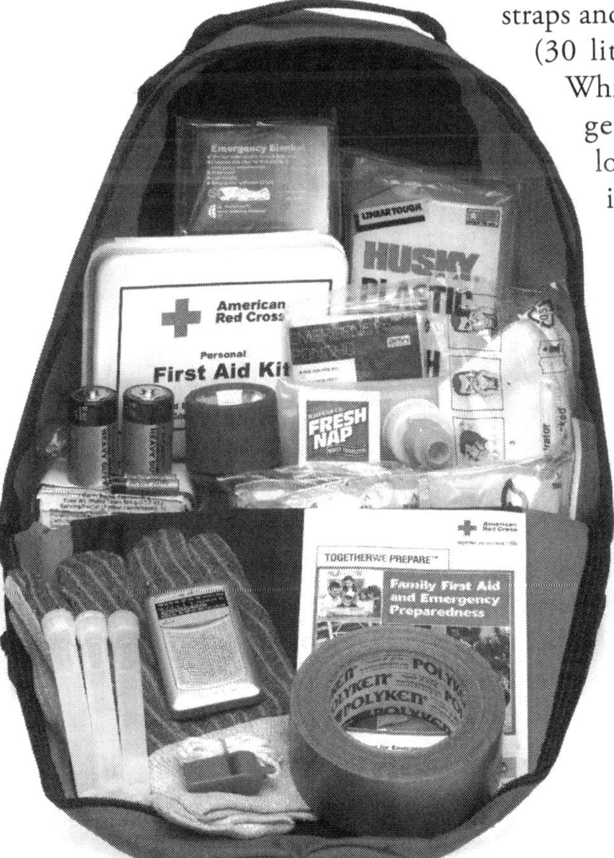

http://www.gofastandlight.com/American-Red-Cross-Personal-Disaster-Survival-Bug-Out-Kit/productinfo/SU-L-PERSKIT/

Now Let's Look at the Contents

Food

Food should be light, convenient, prepackaged and calorie dense. This is not the time to put yourself on a diet. Eat a lot. You'll be burning more calories because of stress, cold, and you may even be bored of waiting, so the more granola-bar type foods you've packed the better. Dried fruit and nuts, nut butters and crackers and energy bars are all good food items for your Bug Out Bag.

Water

Hopefully you'll be able to take some water jugs with you in the car. Remember you'll need a gallon (4 liters) per person per day. If you're bugging out in a vehicle you'll be able to carry three days worth for the whole family. If you don't have a vehicle and you're climbing on a bus to get out of town that's going to weigh a lot and be pretty tough to carry very far.

So having a system to purify untreated water is a good idea. Boiling water is an excellent technique but it's unlikely you'll have a stove or enough fuel to do this on a large scale. But it is an option.

You can purchase portable pump filters. Some models use ceramic filters and some use activated charcoal. When I used to go on canoe trips I took a ceramic filter and never had any problems with the giardia and other pathogens that can make you sick from drinking unfiltered water.

The other option is chemical disinfection. This usually comes in the form of iodine pills. These will kill most of the common pathogens you might pick up from lake or river water. You can buy these pills at some drug or camping supply stores. I used to use these when I backpacked in high school. I found they did affect the taste of the water so I always added "Tang" to cover it up. Back in those days Tang was pretty cool, since, as the commercials suggested, it had been designed for astronauts! Today there are a wide variety of powdered drink crystals that you can use to mask the iodine pill taste!

First Aid Kit

The First Aid Kit in a Bug Out Bag should be fairly simple. If you're evacuating an area because of an impending storm, authorities will probably be directing the evacuation and there will be first responders available. It would be helpful to have a basic kit for cleaning and bandaging cuts and scrapes. For more serious injuries like broken bones, let's assume that professional care would be available. Find a good first aid book and keep it in your bug out bag. This serves two purposes. First, if you do have to put a splint on a broken limb, the book will provide instructions on how to do it. Second, if you're stuck in some emergency shelter for a few days it'll provide you with something to read. Remember the resolution you made to take a First Aid course after reading this book? Well if you didn't get around to doing so, you'll now have a book to teach yourself first aid! How great is that?

Flashlights and headlamp

If you have no other tool to take with you into the unknown, take a headlamp! Light comes in pretty handy during a power outage and having both hands free to complete tasks in the dark is almost as important. City people forget just how dark it gets even inside a blacked out apartment, with no streetlights. Flashlights are great, especially LED ones, but have both. Besides, when you strap on a headlamp you can fulfill that fantasy to be a 'coal-miner' that you've had since you heard Sissy Spacek sing the Loretta Lynn song in the movie "Coal Miner's Daughter."

Cash

I don't think you can say enough about the benefits of having cash around, especially during an emergency. If you find yourself becoming a "cashless" person, relying solely on debit and credit cards, stop it! That's going to end in tears and frustration. Even if the power stays on in your area, bank and computer networks required for all those electronic transactions may be disrupted elsewhere. That local store has bottled water for sale and you want it. No, in this case you really NEED it as opposed to just 'wanting' it. You need a way to conduct that transaction and they probably won't be taking debit cards. Have cash!

Assorted "Handy" Things

Matches and lighters will come in very handy if you're displaced and have access to firewood for warmth. Make sure you've got lots of I.D. so authorities have a way to verify where you're from, especially if you're trying to get back into an area that has been closed to prevent looting. And an extra copy of your emergency plan should be in your Bug Out Bag as well. You're going to have it memorized in no time!

Blankets

If your pack is big enough and you can fit a sleeping bag in, it could be an incredible help in a time of dislocation in the cold. If not, consider an emergency or Mylar blanket, which are small but are designed to reduce heat loss. They stop heat loss through convection, stop evaporation via perspiration and reflect heat back to the person under it, while being wind- and waterproof as well. You won't find them in your honeymoon suite in the Poconos, but they come in handy in a hurricane.

Tools

A multitool is a great device to have in your pack since it can help with so many repairs you may require. Duct tape is a universal tool and while it may not help you hold up walls blown down in a tornado, you'll find it has 1001 (if not 10,001!) other uses in an emergency. A hand crank radio will be helpful to keep in touch with what's going on elsewhere in your area and to give you some idea of when you might be able to expect help. I'm a big fan of the new ball-less whistles which are incredibly loud and could signal your location to others even in the loudest windstorm.

Hygiene Stuff

Toilet paper is bulky but emergency shelters can run out of stuff, and if you've never had to deal with life without toilet paper, you're going to quickly discover what an insane luxury it is! Having a roll or two will be an incredible stress reducer. Wrap them in a watertight, zip lock-type bag that you can take back and forth with you to the bathroom. And following up on that subject, alcohol-based hand sanitizer, whether in individual packets or a bottle, is really important to have. If water is at a premium, it probably won't be available for washing. This is both a safety issue because this is one time you want to reduce the likelihood of picking up some bug via your hands, and it's a mental issue. Going days without washing your hands will affect your state of mind and further deplete your mental capacity at a time when it will be taxed enough.

Even having a toothbrush and toothpaste could be an incredible boost. Yes you can go a long time without brushing your teeth, but having some semblance of normalcy and personal hygiene will go a long way to keep your spirits up. Go ahead, indulge yourself, throw in a toothbrush!

Facemasks

The Red Cross 72-Hour Disaster Preparedness Kit includes 2 of the N95 masks that they identify as "dust masks." N95 masks are a type of disposable respirator. They have been approved by the FDA (Food and Drug Administration) and the NIOSH (National Institute for Occupational

Safety and Health) as a suitable mask to use to help protect the wearer against particulate contaminants. This would be helpful in the event of an earthquake or building collapse where there's a lot of dust around. You see people using less expensive dust masks on those home renovation shows when they're knocking down walls and kicking up a lot of dust. These types of masks are designed to remove the bulk of the dust particles floating around but they are not fool proof in terms of stuff getting to your lungs. Most people don't wear them correctly anyway so while they're a good idea, you should be realistic in your expectations of them. They do not eliminate the risk of contracting any disease or infection. Often when you see people in news reports during pandemics wearing these masks it's as much a courtesy to others so as not to spread germs if they think they may be contagious.

If you are concerned about protecting yourself against disease or infection, and you are able to spend more on your preparations, you should invest in an airtight facemask, or a "gas mask." I'll discuss these briefly in the next chapter

Summary

So there you have it.

You've got your house ready for an emergency, you've got your car ready to go if a crisis hits, and now you've got a Bug Out Bag if an approaching storm suggests it's time to move to higher ground.

So you can relax now! You're ready for the next big thing to come your way. But these are short-term fixes. In Part III I will discuss some of the longer-term issues that will affect us all in the future. These will take a little more time and potentially expense to deal with, but are just as important.

8 Your 'Get Out of Dodge' Bag

Going Down the 'Survivalist' Road

I debated writing this chapter. I consider myself a "prepper," but I'm not a survivalist. Living off-grid and growing most of my own food was a choice I made for environmental and spiritual reasons. I hated the city. I wanted some land. I wanted space. I wanted to generate my own power from renewable sources. I wanted to grow my own food.

If I couldn't leave my house for 6 to 12 months, and no one delivered anything to me, I'd live pretty comfortably. In fact I can see a day in the near future where I'd never have to leave and could exist here comfortably, indefinitely. Of course "comfortably" is a relative word. I would run out of things like coffee and tea pretty quickly, but I'd adapt and be growing the plants for my own homemade "herbal" tea in a few months. Or I'd be finding them in the woods that surround my house.

This ability to live independently for a long time actually puts me into the league of survivalists albeit unintentionally. Independence is their goal. Our reasons for setting this goal are different, but we've sort of arrived at the same place.

I have called this book *"The Sensible Prepper"* because I think it is logical to take some steps to prepare yourself for some physical dislocation in the future. I think it could get bad, but I also think humans are pretty resilient and it's unlikely we'll descend into chaos. That is the fear of most survivalists. As resources diminish, countries will go to war over what's left. As population growth strains our planet's ability to feed everyone, hungry people will start to behave badly. Countries with starving populations with weapons will look at neighbors that still have a surplus of food and decide to take it, or to try. If a river runs through your country first and I'm not getting enough flow to irrigate my crops and feed my people, I'm not going to be happy with you.

Someone with a 'survivalist' point of view will look at me and say I'm just being wishy washy.

If you honestly believe there will be some disruption why wouldn't you just take it to the next step and suggest that there is the distinct possibility that this could all lead to a pretty chaotic future? And it could unfold very quickly. So why wouldn't you plan for such an event?

Well I kind of have even though it was inadvertent. I just sort of stumbled into this state of preparedness. I was doing it to prove I could produce almost all of my own power, and grow almost all of my own food. I took it as challenge. The end result puts me squarely where most survivalists want to be.

Section III is going to take a bigger picture view of what I think you should be doing to prepare for the future. And in most cases what I suggest for a 'sensible prepper' will just be taken to the next step by a survivalist. If I advocate that you should have 6 months food in your root cellar and pantry, a survivalist will suggest that a year's worth is more appropriate. If I suggest it may be prudent to have some form of firearm protection and perhaps 100 rounds of ammunition, they'll suggest you have 5 firearms and 1,000 rounds of ammunition.

I'm not arguing with them, I'm just saying you don't have to start there. And in fact with the number of things you should be doing I don't think it makes sense to put so many of your resources, especially financial ones, into stock piles of guns and ammo. Not that there's necessarily anything wrong with having some guns and ammo, it's just not the place to start.

I also believe that should things go downhill, we won't necessarily experience it as a "fast crash." I think you may start to notice things deteriorating over time, which means you should have time to improve those areas of your preparations that you feel need expanding. In the next section I'll suggest the importance of "liquidity" of your finances, so that you'll have money available to you to capitalize on these situations. The advantage of going through the preliminary steps I advocate is that it will make it easier when you need to bulk up quickly. If you have purchased a firearm, in many jurisdictions you will have had to go through various government regulations. But once you have and you have the documentation you require to buy guns and ammunition, any subsequent purchases will be much easier. So the more steps you can follow in advance with the intention of stocking up at a later date, the better.

So just so you can dip a toe into the survivalist pond, let me outline some of the items a hardcore prepper might have in their "Bug Out Bag." You may see some items and think "Well hey, that's a good idea for mine." And that's great. Just remember when you put a handgun in your bugout bag you ratchet up the intensity and potential consequences of your plan. It may be absolutely crucial for your well being in a situation where you've had to "bug out" because chaos has set in. But if you've just evacuated to an emergency shelter until a hurricane blows over, the authorities coordinating the whole movement of people and resources may not be happy that you have a handgun concealed in your pack. It may be legal in your state, or it may not. But if you're sleeping on a cot in a gymnasium, you may not be able to keep your eye on your backpack all the time. You may let your guard down to get some food or use the facilities. And no one wants a kid finding a handgun in a crowded room.

So I just pass that along for thought. You have to decide whether the situation requires this level of action. Chances are if you need a pack heavily weighted with 'self-defense" items, you'll know things are pretty bad.

Here's a great image I found on the web for a great survivalist "bug out bag."

You can see it in colour at www.uncrate.com/stuff/equipment-bug-out-bag/ or www.apocalypsepak.com/ or www.thispak.com/

Here are the main items that a "Get Out Of Dodge" (GOOD) bug out bag may contain;

- MREs
- freeze-dried dinner kit
- UV water purifier
- ceramic water filter
- Shot gun – ammo
- knife
- hand gun – ammo & holster
- silver coins
- compass
- camping equipment – pot, cutlery
- fishing pole (small)
- hammock
- sleeping bag
- drugs – painkillers, broad-spectrum antibiotic
- maps & compass

Meals Ready to Eat – MREs

An MRE is an individual, self-contained meal in a lightweight package designed for the military. It allows soldiers to eat good meals while in tough conditions. The meals are lightweight because they are dehydrated but offer a dietary sound source of 1,200 calories and nutrients. The kits come with a flameless ration heater which allows you to heat the meal up without having to have a stove or cooking device with you. While they may not be considered 5-star dining caliber they offer a much better alternative to eating canned beans 3 meals a day. MREs allow you to have some lightweight, tasty, nutritionally balanced food when you absolutely need it. When you see how much other stuff you're going to be carrying, having tons of calories to fuel your body is going to be essential.

You can order MREs from these sources:

www.mealkitsupply.com

http://www.survival-warehouse.com

http://www.thereadystore.com/mre

Freeze Dried Food

An alternative or supplement to MREs would be freeze-dried food. Again this food will be lightweight and nutritionally sound, the difference is that it will involve a little preparation on your part. You'll have to add water and then have a heat source for cooking. With a lightweight cookstove freeze dried food is a great choice, but remember you'll also have to carry fuel. So what you make up in lighter food can be offset by more weight for the cooking process.

Companies like "Mountain House" have 3-day emergency kits that would be good for a kit like this, or you may want to order more if you expect you'll be without access to food for a prolonged period. (www.mtnhse.com) Harvest Foodworks also has a great selection of freeze dried food. (www.harvestfoodworks.com)

Portable Cook Stove

There are many types of portable cook stoves available that burn a variety of fuels. Most of us are familiar with propane type stoves but like butane stoves they are a gas, which makes carrying them in a backpack problematic. My preference would be a single burner kerosene or "camp stove fuel" type cooker. The fuel is liquid and you can carry a reasonably sized container with you. You pressurize the stove by pumping air into it then fire it up with a match.

This is a "Sterno Stove" which I talk about in Section III. It is extremely lightweight and an option to consider for this kit.

Portable fishing pole

If you see yourself having access to a lake or stream that might have fish in it, having a small, portable fishing pole would be a good idea. Hopefully this assumes you've had some experience catching and gutting fish, although it is a skill that can be learned if circumstances call for it.

Portable water purifier

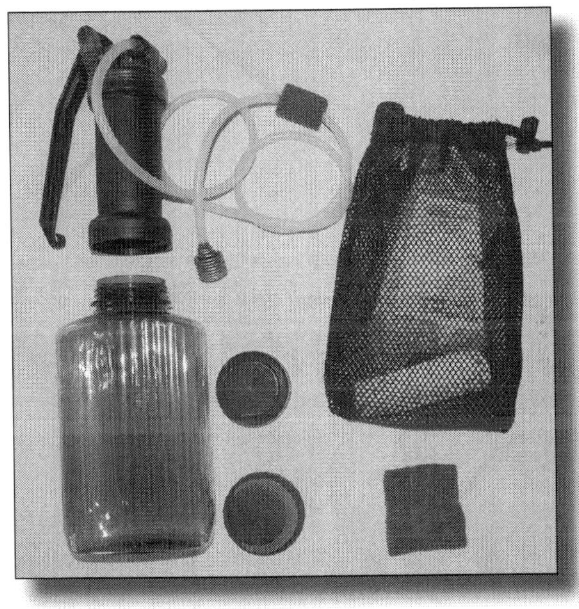

It makes sense to invest in a good water purification system. I used one for many years on canoe trips treating water that was known for having guiardia and other water pathogens that might have made me sick. I'm a real fan of these. They can pump through a lot of water and you can do maintenance on them to keep them working well.

www.cascadedesigns.com

Compass and Maps

Knowing where you are and where you're going can be pretty helpful in an emergency situation. Maps of the area you'll be in are essential, as is a compass. I know what you're saying; "I love my GPS and it works great." It works well when all is well with the systems and the people maintaining the satellites, but remember the type of situation we're discussing here. So let's assume the GPS system may not be operational but you still need to navigate your way. Get a compass and learn to use it. Turn it into an adventure. I remember an orienteering exercise in high school, which started out well enough. The area had a number of ponds, marshes and swamps not well marked on the map. Early in the day we'd take a fix on a spot on the far side of the swamp and walk around the swamp to resume from that point. By the end of the day we were wading through swamps up to our armpits. Now that builds character!

Sleeping equipment

Depending on your conditioning you might be able to include a small tent in your backpack. At least you should have a tarp to give you cover. A hammock will allow you to stay off the ground, crucial if it's cold or wet. And finally a sleeping bag could make the difference between a good night's sleep and a recharged body in the morning, or hypothermia. Stay warm and dry and pace yourself.

Drugs

Having a few basic drugs in your kit will be helpful. A painkiller like hydrocodone or strong acetaminophen (Tylenol) or ibuprofen is a great starting point. While you may not be qualified to diagnose someone fighting an infection, if someone with you was sick enough to have to stay put, then a broad spectrum antibiotic could be critical. Drugs like this require a prescription so hopefully you a good relationship with your doctor and can convince them of why you feel this is important for your "GOOD" bag.

Silver Coins

While I discuss this in greater detail in Chapter 22 having some silver coins may come in very handy. If you're in a situation where you've headed out with your GOOD bag, then you may encounter people you'd like to trade with who won't be accepting paper dollars. Precious metals have always been "real" money and people may be prone to return their use in a crisis. Silver is much cheaper than gold which makes it easier to acquire, and because it generally is worth less than gold it makes it easier for smaller transactions.

Knives

A knife comes in pretty handy in a camping situation. You'll find lots of uses for it and it's pretty low tech, so it doesn't require any maintenance or inputs (like bullets). A good quality knife with a sheath you can use to attach it to your belt is a good idea. The first grizzly bear you run into will run screaming when you pull out a knife.

Handgun

Well there you have it, I've crossed over the line into the survivalist zone suggesting that you carry a knife. Yes, but if you've got a "Get Out Of Dodge" pack on your back, you probably could use one. In a bad situation, hopefully you could use a knife just as a deterrent for self-defense.

And from a self-defense perspective, a handgun may be in order. If you are going to own a handgun, know how to store it properly. And know how to use it. Many people don't operate guns on a regular basis so aren't comfortable with them. If you were in a situation where you needed to use a gun, it would be good if the person you're confronting can see that you know what you're doing. You also need to know the legality of owning a firearm such as a handgun so you need to check with local authorities about the licensing and requirements of purchasing, owning and storing firearms and ammunition.

Shotgun

If you think a handgun is intimidating, a shotgun ratchets up the intimidation factor to "11" (assuming 10 is the highest ranking). Shotguns send out a spray of pellets so you don't have to be accurate and that "chich chich" sound of a shotgun being cocked ready to shoot is easily recognizable by most people. The challenge with a shotgun is simply going to be its weight. It is going to be significant contribution to the weight of your pack, as will the shells that you'll need for it. If you're going to be strapping a shotgun on your back, it will mean that things have gotten pretty out of hand. Lets hope that it never comes to that. However, if you believe that it's a possibility, you should become very comfortable operating your shotgun, and you should probably strap on your "Get Out of Dodge" backpack and go for some full day hikes. This is the only way to determine if you can realistically carry all this equipment over the long haul.

A "Bug Out" Location

Many of the people who are expecting some form of dislocation in the world as we know it have set up a location to bug out to. Some people call them "cottages." While a cottage on a lake sounds like a lovely place to escape from city strife, it may not be an ideal location to "bug out" to, at least not over the long term. The challenge will be its location and the property. Areas that are best suited for cottages, like around lakes, are very sought after and therefore quite expensive. So the lots tend to be small. For a temporary refuge they will be fine. But for the long term what you need is land. If you are going to be there for a while you need to be able to grow food. Ideally it would be great to have a woodlot so you can harvest your own firewood for heat and cooking too.

In Section III I'm going to make suggestions for where you might want to live. Some people have accepted that it is difficult to earn an income far from an urban center, so they continue to live near a city but have a rural location to "bug out" to. For many this will be a luxury. Having a second home along with its taxes and upkeep will just be out of many people's reach. One way to keep the costs down would be to just buy property and have camping supplies there ready in case you need them. Of course this works better if you're there in August rather than January.

If you do end up with a "bug out" location, you need to do the math on how you're going to get there. The news from Hurricane Sandy quickly switched from destroyed homes to hour-long lineups for gas, if you could get it. You need to keep your gas tank as full as possible and have enough gas in jerry cans to get you to your "home away from home. " And in case you run out of gas or have trouble getting to your bug out location, your "Get Out of Dodge" backpacks will be perfect for that long hike. These may seem like crazy concepts. But watching New Yorkers gathered around a tiny generator several days after Super Storm Sandy hoping to eek out a few luxuries that electricity provides, reminded me of how quickly the veil of normalcy can fall from a technologically-advanced society.

And on that note, lets move on to Section III. I think with many of the suggestions I make in the big picture view you may not require a Get Out Of Dodge bag. You should be fine to stay exactly where you are, which I think would be the goal for most of us. So let's start examining steps you can start taking now to make that a reality.

9 The Contagion Factor

First off I have a homework assignment for you. A fun one, before we start this section. One that involves watching a movie! Woo Hoo!

Find a way to watch the movie "Contagion," that was released in 2011. It's a great movie with lots of big name actors like Kate Winslet, Matt Damon, Gwyneth Paltrow, Jude Law, Laurence Fishburne … you name it, and they're in it!

I want you to watch this movie as a motivator to get you working on prepping. There's no use reading this book and other information about prepping and not taking action. I want you to take action and I think this movie will help to motivate you, while at the same time entertaining you.

The movie follows the breakout of a global pandemic and its effect on society. They don't identify the cause of the pandemic, it's just one of those nasty bugs that happens when viruses jump from one species of animal to another species of animal and then to humans. This happens all the time. Many flu bugs are spread this way. Swine flu originates in pigs, avian flu originates in poultry, and where people live in close proximity to their animals, it is easily spread.

We have had pandemics in the past. The SARS outbreak (severe acute respiratory syndrome) in 2003 killed 800 people worldwide but luckily authorities managed to stop it from becoming too widespread. Many people were critical of the hype surrounding 2011's "H1N1" flu when the World Health Organization (WHO) seemed to over react about its potential to become more lethal. We know that pandemics are a reality.

You might have heard of the 1918 "Spanish Flu" pandemic that killed up to 50 million people of the 500 million it is estimated to have infected. Some historians have suggested that the close proximity of troops in World War I and their subsequent disbursement back to their homelands increased the transmission of the flu. Most flu outbreaks usually kill the very young and the very old or those with weakened immune systems, but the Spanish Flu was different and predominantly killed healthy young adults.

So we know that a pandemic can kill a lot of people in a short period of time, including healthy people. Fifty million people in 1918, when the population of the planet was only roughly 1.8 billion, was a lot. This was at a time when the movement of people was much more restricted. There weren't many cars. There wasn't air travel. It would have taken an American soldier months to get back to America from the front lines by taking a slow ride on an ocean going ship.

The 2014/15 outbreak of Ebola was in large part confined to Africa, but isolated cases have arrived in North America and health authorities have dealt with them very aggressively because of the hazard diseases like this pose.

Today we have huge numbers of people flying all over the world, so it is much easier for a pandemic to spread at a much greater speed. We also have 7 billion people living on an increasingly overcrowded planet, and many of them live close to and interact with animals.

I pass this information along not to make you paranoid or to try and scare you; I just think it's important for everyone to think about risk. The risk of you being killed by a pandemic is small. The risk of you being impacted by one is much larger.

That's what I like about the movie "Contagion." It takes a logical and realistic look at how authorities would deal with a modern day outbreak as virulent as the Spanish Flu. Once it has been determined how a disease is being spread, governments would consider restricting travel. Air travel might be suspended. As the pandemic spread from one community to the next, individual jurisdictions would have to look at how to deal with it. If your city or state didn't have any reported cases but your neighboring jurisdictions did, the government might be inclined to stop the movement of people and goods to protect their electorate. This would involve quarantine. This would require borders to be closed. It would involve people's movement being restricted.

There are many in the "prepper" movement who might take umbrage at this, the government trying to control people. But realistically, if there is no cure and no vaccine yet developed, doesn't it simply make sense to try and stop it from spreading?

If travel is restricted, think about the supply systems that keep our society functioning. Food would be an important consideration. Trucks supply the food we buy at the grocery store. Much of that food comes from quite a distance and crosses many state/provincial borders to get to us. So we can assume that the supply of food will be affected. And, of course, once people became aware of a limited supply, they would try to stockpile food quickly and stores would soon become emptied.

There is a trend in the industrial world called "Just In Time" delivery. In the past, manufacturers kept huge inventories of parts ready to assemble into finished products, like cars. Warehousing all of these parts was seen as an unnecessary expense so manufacturers began to insist that parts be delivered as they were needed, or "just in time." This practice might make sense when manufacturing cars but it doesn't make sense for a country's food supply. Unfortunately though, that is very much what our modern food infrastructure has become… a "just in time" delivery system that gets food to the grocery store just in time. And when trucks stop rolling the inventory can drop fast, potentially precipitously.

That's what you see happening in the movie. Grocery stores are quickly emptied. After they've looted all of the grocery stores, hungry people begin looking at homes as their next target.

Now think about the infrastructure that you rely on to keep you warm, to keep your lights on, your freezer cold, your toilet flushing and all of those other creature comforts you take for granted. What happens when employees at the electricity generation station get sick? Or at the municipal water supply? Or members of the police and fire departments? What happens when there's no one left to pick up the garbage?

Again, I am not trying to be hysterical; I am just encouraging you to give some thought to this scenario.

You need to ask yourself, "How would I be affected by an event like this? How comfortable would I be? How long would my food hold out? How long would I have water to drink? How would I keep clean?" Asking these questions doesn't turn you into some kind of crazed militant. It's just a logical human exercise to go through.

I always like to draw on the analogy of the 1998 "Ice Storm" that affected the northern United States and a large part of Ontario and Quebec in Canada. Millions of people were without electricity for weeks. And things got very bad for many people very quickly. They couldn't heat their homes. They had no water. They couldn't prepare food. Pipes froze. The military had to be called in.

And yet a few generations ago people would have been unaffected by an icestorm. It would have been an inconvenience, but it wouldn't have been a crisis. So let's make sure you can ride out the storm and not have to be the person needing a rescue.

I like to tell the story of friends who live off the grid near Ottawa, Ontario and were in the heart of the ice storm of January 1998. The military was evacuating people to shelters because they were freezing in their homes and had nothing to eat. Water pipes were freezing and bursting in homes without heat. It was not a pretty scene. The military went door-to-door checking at each home. When they arrived at Bill and Lorraine's house, Bill opened the door in his bathrobe because he had just had a hot shower. Heat from their woodstove wafted through the house, Mozart played on their stereo and Bill had a steaming cup of cappuccino in his hand, something he is rarely without. The two soldiers looked at their clipboard, looked at Bill, heard the music, felt the heat, smelled the coffee and said, "Well, we think we know the answer, but is everyone in the house alright?"

"Never better!" was Bill's reply, and this should be your mantra when you've finished reading this book.

PART III

Sensible Prepping The Long Term/ Big Picture Approach

10 Introduction to Resilience

The future isn't going to look like the past.

The first two sections of this book examined some of the challenges that the world is facing and how they may impact you and your family. I've suggested steps that you can take to deal with the clear and present danger of an immediate risk to you and your family.

It's time to make yourself more resilient. By this I mean having a strategy to deal with disruptions and dislocations that may not pose an immediate danger but which over time could have an affect on your health and well being. Climate change, economic collapse, peak oil and resource depletion; these type of challenges will have an impact over longer time periods but could pose just as big a danger as an approaching hurricane. Sometimes it's these slow moving dangers that are easiest to ignore.

The following chapters are a road map to guide you on your way towards resilience. They look at all the major "needs" that humans have. These needs are located on the lower rungs of Maslow's Hierarchy of Needs which I discussed with the graphic in Chapter 4; the need to eat, the need to stay warm, and the need to stay safe. The future will require many of us to spend much more time and energy dealing with these basic needs. But this is not to say that you won't also be involved with the higher rungs at the same time. In fact, you may find that dealing with basic human needs brings a new joy and sense of fulfillment to your life and that those higher-level needs start to look after themselves.

I would suggest as part of this longer time horizon you start developing a "Five-Year Plan." Take the information provided here and chart a course for your next five years. Set a goal, whether it's to buy a smaller house closer to work so you can take transit or to buy a place in the country so that you can grow all your own food. Figure out what it's going to take to get there, and then develop a plan. Regularly put money into the "Five-Year Plan Account." Monitor your progress to help you stay motivated.

There are numerous reasons to embark on a program to make yourself more energy, food, and financially independent. This book will point them out as we go but will keep coming back to the main benefits: it's good for your health, good for the health of the planet, and good for your resilience to a less certain future.

11 Where to Live

Where you like to live is a very personal thing. Lots of things affect it. Where you work. If you want to be close to family. If you like the convenience of living in a city, or if you long for the wide open spaces of the country. Some people like to be around people and freak out when they find themselves isolated. Others are almost allergic to the crowds. John Cougar Mellencamp "can breathe in a small town," but Frank Sinatra sang "If I can make it there, I'll make it anywhere, it's up to you, New York, New York." It takes all kinds.

Most North Americans choose to live in and around cities. In 1950 about 60% of North Americans elected to live in urban centers. By 2005 that had grown to over 80%. Over the past century more and more people have chosen to move from the country to urban centers. They've been drawn by jobs, cultural pursuits, and, of course, $5 lattes at Starbucks.

Cities can be very sustainable places to live. Lots of people hear of someone living off the electricity grid in the country and make the assumption that this is the definition of sustainability. If they live in a place where lots of the electricity is generated from coal, there's no doubt that the carbon footprint of that off-grid dweller is going to be much smaller when it comes to electricity. But how do those people earn their living? Are they retired, or do they own a home-based business? If they don't have to leave their rural home very often, then they're doing the right thing for the planet.

What if they work in town though? Suddenly their half-hour or one-hour commute into an urban area has negated the benefit of their solar panels. And what do they use for their sources of heat? If they have simply shifted all of their heavy electricity loads, namely heat sources, to propane, then they really haven't changed their energy footprint significantly. In fact, someone living in a city and using electricity from a nuclear plant for their cooking and hot water is probably further ahead in terms of the CO_2 that they're putting into the atmosphere.

What if that urban dweller also uses public transit to get to work rather than driving as so many rural folk do? And what if they live in a condo or apartment rather than a detached home? Think of the heat loss in an apartment. Generally, you only have the outside wall exposed to the elements. Three of your four walls abut other inside walls, so you don't have the heat loss you would have in a detached house where every wall is exposed to cold outside air.

Someone living in an apartment in a city, taking transit to work, using energy as efficiently as they can, and shopping at the local farmers' market has a much smaller footprint on the planet than many off-gridders. I'm not saying one is better or worse, just that there are many shades of gray in this debate. The key is to determine how you can reduce your impact.

When it comes to dealing with the challenging times we're confronting, where you live is going to have a major impact on how you deal with the challenges. There is no right answer; each has advantages and disadvantages. First let's look at various regions of the country and then consider the advantages and disadvantages of rural and city life.

Step Away From That Coast!

Climate change is happening and it's happening much faster than scientists had predicted. I believe therefore, that you should try and locate away from a coast, as high above sea level as possible.

This might sound like a crazy, bordering on impossible, suggestion. A huge percentage of Americans live near the coasts. Work your way around the country... Seattle, Portland, San Francisco, Los Angeles, San Diego, Houston, New Orleans, Miami, Washington D.C., Baltimore, Philadelphia, New York, Boston. It's kind of scary.

Sea levels are rising for two reasons. One is the melting of glaciers and ice in the Arctic. Melting glaciers have a double whammy effect. As more ice melts there is more darkly colored water to absorb heat and less light-colored snow and ice to reflect the solar radiation back out into space. As more ice melts, the conditions are created to accelerate the process. The second is thermal expansion. Hot water takes up more space than cold water. So as the ocean warms, it takes up more space.

This effect was well illustrated during Hurricane/Superstorm Sandy. The storm surge, water blown onto shore by high winds, was compounded by the fact that the sea level has been rising, and the ocean was warm which resulted in increased destruction. Those areas in New Jersey and New York like Long Island that experienced the worst of the storm damage are also places most at risk because of rising sea levels. As I watched the reconstruction of New Orleans after Hurricane Katrina I wondered how wise it is to undertake such a task in an area that is effectively at or below sea level, as sea levels are beginning a steady increase. And now we're doing it again after Superstorm Sandy.

I realize this sounds harsh but you have to accept the fact that sea levels will continue to rise for the foreseeable future. Humans are showing no sign of reducing our consumption of fossil fuels, and as we pump more carbon dioxide into the atmosphere the impacts of climate change will accelerate.

The next question becomes, "How far 'inland' do you want to go and how high above sea level should you go?" Data from previous periods when the earth was much warmer than it is today suggests that sea levels could rise a lot. There are always going to be places around the coast that are higher than others. Portland and Seattle have areas close to sea level, but they have lots of areas with higher elevations. So if you live near a city close to sea level, at least make sure you're as high as you can get. But remember that part of the challenge will not just be rising waters, it will be more intense storms brought on by a warmer climate. While your home may not be prone to flooding it will more likely be in the path of stronger storms.

There is no magic solution to this problem. Many of us are tied to large coastal cities by our jobs. Or family members. But you should realize that over time you are more likely to suffer some of the worst consequences of a chaotic climate. Extreme weather events will become more frequent and increase in intensity.

As you start to look to moving inland, remember to factor in other potentially dangerous weather events. You might want to avoid moving to "Tornado Alley" for example. This area is roughly defined as the areas in between the Rocky Mountains and the Appalachian Mountains, where tornadoes are most frequent. You can find information on areas most prone to tornados from various sources, such as www.wikipedia.org/wiki/Tornado_climatology. If you are choosing a new location based on avoiding catastrophic climate change events, it would be recommended that you try and avoid these areas, or at least make sure you purchase or build as strong a home as possible. I discuss more specific ideas for tornado-prone areas later in this chapter.

Warmer air holds more water which means that rain storms will become more intense. While one area of the country may be experiencing a brutal drought, others may be getting damaging rain. There are certain areas of the country that are more prone to flooding, but crazy weather can extend these flood-prone areas. So before you buy that dream home in the country, research the chances of flooding in the area. If your real estate agent says "Well this is in the 'hundred year' flood zone, but we don't ever expect to see one of those," run screaming. "Hundred year floods" are becoming regular occurrences throughout the world.

Do your research and have some idea of the odds of flooding in your area. The following websites are good sources of information;

www.noaanews.noaa.gov/stories2010/images/usfloodrisk_spring2010.png

www.weather.com/life/safety/flood/article/flood-are-you-at-risk_2011-10-19

While we are on the topic of water we need to talk about areas that are facing a lack of it. Much of the southern U.S. is water challenged today. Climate change models suggest this is going to get worse and many drought-prone areas today will become desert like in the future. You need water to drink, to bathe and to grow food. If you are considering a move to a southern state, you should think twice. Don't get me wrong, I have nothing against the southern states and I often envy their climate, especially during one of our cold northern winters, but if they are water challenged now, it will only get worse, potentially much worse, in the future.

I know what you're saying right about now. "Well that just about eliminates the entire country!" Well, not really. I would recommend that as the climate warms you'd be better off if you live in one of the northern states. You may just want to make sure that you're not too close to the coast or a river that tends to flood. In 2011 the Missouri River flooded areas of North & South Dakota, Iowa, Nebraska, Missouri and Kansas. You can look up historical records to see if your prospective home is in an area that may be affected by events like this. The 2010 Tennessee flood that flooded downtown Nashville and the Grand Ole Opry was called a "thousand year flood." While I suppose you could convince yourself that it won't happen again for another thousand years, I would be more inclined to seek higher ground.

If I were looking for to relocate today I'd just keep humming Stevie Wonder's song, "Gonna keep on tryin' 'til I reach the higher ground."

Country Mouse/City Mouse

Once you have determined the area of the country that you're going to locate to, let's look at the two main options.

The two best choices are living in the city or living in the country. The wrong choice is living in the suburbs. The suburbs will become less and less desirable as many of the challenges converge on us. Suburbs were built using a model of urban planning that is premised on large amounts of cheap energy. The homes are far apart, and they are located far from shopping and jobs. You have to drive everywhere you want and need to go. Since the homes are so spread out, it is difficult for local governments to service them properly with transit. Transit works best when people live densely, as in a city. The suburbs are so spread out and thinly populated that it is difficult to provide good transit services. That leaves you reliant on a car, and with peak oil driving the cost of fuel up from now on, this is going to get very expensive.

Suburbs tend to have single-detached, large, and hard-to-heat homes, making them the least desirable place to live in the future. The infrastructure in the suburbs will be an issue. For every mile of sewer or water line, far fewer residents are serviced in the suburbs, so a city government with diminishing revenue is apt to put money where it will do the most good, and that's in the more densely populated city areas, not the suburbs. If you live in the suburbs now, perhaps you should be planning a move in your Five-Year Plan, either to the city or to the country.

Living in the City

Pros

The great advantage of living in the city is that many of us already live there. It means less disruption and more continuity. If we've got lots to deal with in terms of the economy and our jobs and food and energy issues, it's better that we don't also have to deal with the jarring impact of a move to a radically different way of life. People accustomed to having a myriad of shopping and services close at hand often have trouble adjusting to life in the country where that level of service just doesn't exist.

With most citizens living in cities, governments at all levels will have a natural tendency to devote resources to those areas. Governments will be pulled in many directions as multiple challenges converge on their resources, but they're going to have to make sure they keep the majority of their citizens happy, especially as long as North American governments are democratically elected. It only makes sense to keep the majority happy if you want to get re-elected.

The other advantage of living in the city is its proximity to jobs, or potential jobs. Businesses tend to cluster in urban areas to be close to a work force and services such as electricity, water, sewers, and high-speed Internet.

So if you have a job, it's likely already in an urban area. If you are looking for work, the greatest number of potential employers is in the city. If your job is accessible by transit, this is even better.

Transit is going to be one of the priorities for governments, In fact, as the world starts to become aware of the implications of declining oil production, governments will finally start to make the investments in transit that they should have been making all along. In the spring of 2008, as the price of oil was approaching $150 a barrel, transit ridership throughout North America increased dramatically. If the price of oil had stayed there much longer, transit authorities would have had to begin a crash program of infrastructure investment to add the trains, buses, and subway cars required to move the higher number of riders.

Oil has returned to levels around $100/barrel and soon the market will return to its recognition of just how precious the remaining oil in the ground is, and the days of happy motoring will be over. Transit will become the most cost effective option for a large part of the population, and you'll find the best transit in the cities.

As Russia went through its jarring transition from communism to capitalism, one of the things that allowed life to go on as usual was the fact that most Russians didn't own a car. They were used to public transit as their major mode of personal transportation, and even with the problems that befell the transitional governments in Russia they were able to keep the buses and trains moving, which kept people working and the system functioning.

You should expect there to be jarring impacts in the North American car-based transportation model, but I believe you will be able to count on the government to keep public transportation moving. This means you'll be able to get to your job and to go and buy food.

Food is going to become an increasingly bigger part of your future. It's going to get more expensive and it's going to be harder to come by.

There are so many fossil fuel inputs into our food supply, thatthe rising cost of energy is going to hit the price of food in a variety of ways. In the city, you are pretty much completely dependent on someone else to provide your food. You can certainly have a garden if you own your own home, and you can practise intense gardening to try and squeeze every ounce of food out of the soil, but the reality is that in most urban areas you simply won't have enough land to make a huge dent in your food budget. You are still going to need to buy food from someone else.

If our current food infrastructure remains relatively intact, stores will still have a lot of food, but the variety may be greatly reduced. You may have to change your approach to eating and substitute food choices you wouldn't have considered in the past. Heavily processed foods will likely be proportionately more expensive, because all that processing requires energy; so you'll likely need to start eating lower on the food chain, meaning fewer reconstituted potato chips in a cardboard tube and more potatoes. And the most economical source of those potatoes in a carbon-constrained future will be local farmers.

Local farmers are going to want to maximize the return on their investment of time and money, especially their fuel costs. What they'll probably realize is that it makes much more sense for them to market their food in areas with the highest population density, and that will be cities. Farmers' markets are becoming common fixtures in urban areas and they are the beginning of a model of food distribution that is going to become the norm. Farmers are going to drive to an area of high density, and people in that city will have to find their way to that market, be it on foot, bicycle, or transit. This form of transportation will reduce how much can be carried, so people will be eating fresher food that they purchase more often.

So living in a city and using a farmers' market is going to be better for you by improving the quality of the food you eat; it's going to be better for the planet, reducing the miles that food travels before it gets to your plate; and it's going to help you minimize the effect of rapidly rising energy costs.

It is surprising to learn that people who live in cities are much healthier than people who live in the country or the suburbs. People in the outlying areas too often substitute liquid hydro-carbons (gas, diesel) for human calories. They drive more for work and shopping, and even though they have the opportunity to get exercise by cutting large lawns or cutting firewood they tend to use ride-on lawnmowers rather than push mowers, and miss the opportunity for personal exertion. City people know the hassle of driving. Traffic makes moving around a city slow, and parking adds to the expense and inconvenience of a trip. Taking transit often makes more sense, but the bus or subway stop isn't usually right outside the door, so you tend to walk more. If your favorite coffee shop is four blocks away, it doesn't make sense to drive or take transit, so you put on some comfortable shoes and get walking.

There are lots of other advantages to city life. You'll have more access to cultural events, especially if the economic downturn reduces advertising revenue and we see our 500-channel cable TV world start to contract. You can count on there being lots of creative people in cities who are happy to use entertainment as a source of income. You'll be closer to stores that will help you save money, like second-hand stores and pawnshops, and you'll be able to find a library nearby with an almost infinite amount of free entertainment available in the pages of its books. If you're reading this book on loan from a library you already understand this concept. If you purchased our book, we thank you.

Cities will always be vibrant places to live, with a variety of creative people teaming up to use their talents and resources to deal with the coming challenges.

Pros of City Living - Summary

- Public transportation infrastructure a priority
- Ability to walk and cycle for work and food
- More economic incentives for farmers to bring food to areas with denser populations
- Access to cultural activities

Cons

Living in the city will have some potential downsides. One of those is food. While I think it will be a good place to live if farmers can get their produce into a central market area, what if they can't? What if the effects of peak oil are more dramatic than we anticipate and farmers simply don't have the diesel to put in their trucks to bring their food to the city? If we get to the point where farmers have to go back to relying on horses, this is going to pose a problem.

In the old days, a farmer could hook up his wagon and set out for town with his food to sell or trade. The problem now is that suburbs have made the area around cities a wasteland of strip malls and super highways. Lots of good farmland got chewed up in this process, but it also created a fairly cumbersome barrier for someone from a rural area trying to get to the downtown area without the use of a fossil-fuel-powered vehicle. A horse and wagon just couldn't realistically do it. They certainly can't get there and back in one day and we don't have facilities in cities for horses anymore. So if we reach the point where farmers can't get food downtown to the farmers' market, you'd better have a Plan B that includes you living closer to the farmer.

In the city you are very dependent on other people, not only for your food but also your electricity, your heat, your water—everything that makcs your life convenient. When we started our migration to urban areas, we began giving up our independence and trading our labor for an income based wage that gave us enough money for life's necessities, like food and water and heat. This arrangement has worked for decades, and cheap and abundant energy has made it work well. Power utilities can build power stations and source fuel (coal, natural gas, uranium) that allows them to produce electricity cheaply enough that it's a fairly small part out of most people's paychecks.

Other utilities have been able to purchase natural gas and pump it through pipelines to heat our homes and apartments and businesses. As long as it is cheap they can make a profit and we don't chew up our whole income staying warm. The city can purchase electricity at such a low cost, along with the chemicals needed to process your water and supply it to you at such a low rate, that most people hardly notice things like water and sewer services in their monthly budget.

This model of trading your labor for income to purchase these items works as long as products and services are inexpensive and as long as energy prices stay low enough that there's room for profit. The model goes out the window as energy prices rise and the costs of building and maintaining the infrastructure to keep you supplied with these essentials gets too high. This is going to be the model for the future. A declining amount of income for many North Americans will occur at the same time as the costs of many basics rise dramatically.

People in the country have some of the same problems, but they are more likely to have control of the necessities that city dwellers have farmed out to someone else. Country people generally have to provide their own water from a well. They know how to get the water out and since it usually depends on electricity many already have a backup system like a generator to keep it flowing even if the local utility company stops providing power. In a city you are completely dependent on essentials that you have no control over in either their delivery or their price.

As governments try to adjust to this new environment of rising energy costs and declining tax revenues (as economic activity declines), they are going to have to reduce some or many of the support services they currently provide. Many of the people who receive this assistance live in cities: people who are unemployed, on welfare, or on disability pensions. As governments have to make choices about how to spend money, this group may find themselves receiving less support from the government just as the cost of living is rising because of energy prices. Some of these people are not going to be happy about it. In the last 20 years we have seen a huge increase in the gap between rich and poor Americans.

As more and more people join the ranks of the poor and the gap between them and the rich increases, and as the government cuts back on social programs, they're going to be increasingly frustrated. This will be manifested in the form of increased crime and civil unrest. The poor will look for someone to blame and it will either be governments or authority figures or those who seem to still be doing well. This is basic human behavior. People want to see some degree of fairness. The last 100-year bonanza of cheap energy and abundant economic activity to support government is about to end, and for many it will be a hard pill to swallow.

Some will act out of anger and some out of desperation. We can all appreciate how a person might feel and behave if they or their children are hungry or cold. If they can't get help from the government, they will search for a solution somewhere and resolve it, potentially through illegal means.

This just means that cities are going to become more dangerous places to be. Over the last 20 years we've seen quite dramatic decreases in the crime rate in many major North American cities. I believe the days of this trend are over. There are going to be increasingly angry and desperate people committing crimes, and police departments are going to be stretched to their limits and unable to stop much of it. Ideally their budgets should increase to deal with this, but they'll in fact be going in the opposite direction as cities grapple with less money.

While many "doomers" would suggest that urban centers will soon resemble a *Mad Max* movie, I believe we'll have the resources to keep them quite livable, but living there will bring us into much closer proximity to people who are at their wits' end and willing to do whatever it takes to get by.

Cons of City Living - Summary

- Food harder to come by
- Increased dependence on someone else for necessities such as water, electricity, food, and heat
- Social unrest more likely as income gaps widen and economies have less excess to help those less privileged

Living in the Country

Pros

For many, the first response to “prepping” is to want to “head for the hills.” Here are some of the reasons that may not necessarily be such a bad idea.

One of the major downsides to urban living is going to be a huge advantage of country living, and that’s food and your access to it. Chapter 15 is devoted to the subject of food, so I won’t dwell on it now. The executive summary is simply that food is going to require a much greater effort on your part and it’s going to be easier to grow or access food in rural areas. Most rural homes have more property and hence more room to grow your own food. You also tend to have access to some of the important elements required for healthy gardening, such as manure. Commercial fossil-fuel-based fertilizers are going to be expensive and hard to come by, so finding natural fertilizer is important. As energy becomes more expensive, finding it locally is going to be even more crucial. And if you can’t find it on a farm nearby, you might choose to keep animals and make it yourself.

A well-managed small-scale farm that includes livestock like horses, cows, pigs, and chickens forms an excellent basis for a self-sufficient rural oasis. You can feed the animals on grasses and hay on land that may not be prime for growing vegetables and then use their manure to enhance your vegetable garden and increase its production. Most of us have no background in growing food because we were born and raised in a time when cheap energy allowed farmers to grow far more food than they needed and sell the surplus to us. For us, food comes from grocery stores and restaurants.

Well, that simply isn’t the case. Food comes from the soil. It has to be planted and nurtured and weeded and watered and harvested and stored and canned and frozen and it requires an enormous amount of time and effort. It’s incredibly rewarding, but for most of us it’s a foreign concept that we're about to become intimately aware of, very soon.

If we do live in the country and find a way to earn an income and choose not to grow our own food, we will be in close proximity to people who have food to trade, namely farmers. Farmers have the skills to grow a surplus of food and so will be willing and able to sell or trade some of that extra to their neighbors. It is much easier to trade with a neighbor than to have to transport it to an urban area farther away. The price may not necessarily be significantly less than you'd pay in the city, because the farmer knows that people in urban areas may be able to and have to pay more for essentials like food. But the odds are that some of your neighbors will have food to trade.

Another advantage of living in the country comes from the trend we see now, which is that rural residents are more likely to adopt renewable energy than urban dwellers. There are lots of reasons for this. People in the country like to be more independent and solar and wind power allows this. Electricity can be more problematic and power interruptions more common in the country, so having a system to keep your appliances working during an outage is top of mind with many country people. This happens out of necessity for most. To pump water, you need electricity. So when the power goes out, the water stops flowing. A human can go only 72 hours without water, and for most of us, having a toilet that flushes is pretty important so it's good to keep the water flowing all the time.

People in the country are often more familiar with flooding than city people. Without storm sewers, rural basements can have a tendency to flood after big rainstorms or during the snowmelt in the spring. The way to deal with this is to have a "sump well" where the water collects and a "sump pump" to pump that water out of the basement. The sump pump will cycle on and off as the water rises, drawing the water down and shutting itself off. This pumping can last days or weeks until the water stops flowing into the sump well. As long as the electricity keeps flowing, the water keeps being pumped out. But as soon as the power goes out, the water starts backing up and the basement gets flooded. Rural people who have had to replace carpet or drywall once usually vow that it won't happen again and they make sure they have a backup system to keep that pump on. This can be a backup generator or a renewable energy system with battery backup.

Having more room on their property seems to inspire more people in the country to want to put up solar panels or a wind turbine. While you don't need to put solar panels on a tracker, for many the ability to track the sun as it moves across the sky adds to the attraction of solar power, and a tracker needs a wide swath of unencumbered horizon. There's no reason for many urban dwellers not to put solar panels on their roof, but the greatest number of early adopters have been rural dwellers. Wind turbines require a large unrestricted flow of air to be most effective, so the turbulence created by obstructions in urban areas make them a bad option for wind power in the city. The more open unencumbered areas usually found around farmers' fields and grasslands make wind much more attractive, and rural homeowners often make wind power part of their energy systems. So living in the country seems to inspire and be conducive to homeowners integrating renewable energy into their living arrangements.

I cannot emphasize enough the intimate relationship most rural residents have with water. No matter where you live you need to drink water, and it's very nice to have water to flush the toilet, wash clothes and dishes, and keep yourself clean; but city folk often take it very much for granted. Someone else provides it and the assumption is that it's limitless and safe. Having a well for your water makes you acutely aware of the finite nature of water and the difficulty of ensuring that it's safe. Wells replenish themselves only so quickly, so you have to be aware of how much water is available to you. And with potential agricultural runoff or industrial pollution upstream somewhere in your watershed, you need to have your water tested to ensure it's safe.

With this awareness of water comes the knowledge of the impact of not having it available, whether through loss of electricity to pump it or drought restricting its flow. Country folk tend to have a "Plan B" for what happens when the power goes off or the water isn't clean enough. While I would encourage urban dwellers to be aware of the precarious nature of their water supply, especially during power disruptions, it is much tougher since someone else controls it.

That increased control over the essential operations of a household also applies to heating. Many rural residents are now discovering ground source heat pumps as an excellent and very environmentally responsible way to heat. While it's also an excellent option for a city home, many of the first urban heat pump users had large properties that made a horizontal loop system possible. This will be discussed in more depth in the heating section.

Heating with firewood is also more common in the country. Some people have enough property and a large enough woodlot that they can supply themselves with their own firewood. Those without their own woodlot can usually find local suppliers. Providing you are using an EPA-certified woodstove and burn your wood properly, heating with wood is an excellent, carbon-neutral way to heat and is much more common for those living closer to the source, that being forests. As home heating oil, natural gas and other forms of heating become increasingly expensive, wood will be an excellent option for many people.

Pros of Country Living – Summary

- Ability to grow your own food
- Proximity to farmers to trade and barter with
- Easier energy independence:
  - wind turbines and solar trackers easier to install
  - water you can pump yourself
  - firewood you can harvest or buy locally

Cons

In all things, nothing is perfect, and country life is no exception. The single biggest downside to living in the country is transportation. People in the country drive more. Whether it's to the store, to work, to the restaurant in town for breakfast out, or to the dump, you tend to get there in a car. In a carbon-constrained future, this is going to become increasingly expensive. In a severely restricted post peak-oil world, it could become more than expensive; it could become unaffordable for many. The allure of country life for a great many people has been premised on mobility that cheap oil has provided. Those days are over, so just how attractive the dream of country life is will depend on a number of factors. Do you have to drive to a job to pay your mortgage? How much food do you plan on growing yourself and how much will you have to purchase at the local store? How social are you? How old are your kids? If kids are in soccer and hockey, which entails once-a-week practices and twice-a-week games, how much driving are you going to have to do?

One of the best options would be to have property just outside of a small town, where you could walk or cycle into town, but this isn't always available. It also may be more expensive than property further from town, because it is a more attractive place to live. So where you end up will depend on your budget. And if your budget limitations see you living further away from the nearest trading location, you may be going there less often than you'd like.

Not getting into town very often could be one of the major drawbacks of living in the country. Most of us are acclimatized to interaction with other humans. It's one of the reasons we live in cities. So finding yourself more isolated in the country can be a major drawback for many people. Lots of people try the country thing but end up back in the city to be around people and the other amenities that come with city life.

This may be more of a disadvantage in the future as we move from a money economy in which we trade our labor for dollars which we in turn purchase things with, to one in which we increasingly trade our labor or the fruits of it for the fruits of someone else's labor. This is based more on a barter model which, with collapsing currencies, rapid inflation or the deflation of paper currencies, could become much more attractive and common. While it may not give you the variety we have in our current economic model, it may give much better value. It will also force you to get to know your neighbors better to increase the pool of potential goods and services to trade for. This will happen in the country and in the city, but there's a good chance in the city you'll have a larger potential trading base. The question is whether there are products you need in the city. It is likely that there will more trade in food and basics in the country, while the city may offer more finished goods or luxury items.

Cons of Country Living - Summary

- Transportation a potential problem
  - more driving (or longer distances on your bike or horse)
- Isolation - fewer people to trade with

Communal Living

Some people have looked at other models of living arrangements and have explored intentional communities. While there are no hard and fast rules for this, the model is often based on a fairly large piece of land and a building or buildings which are jointly owned by a number of people. Community members have their own living space and often share a large common area such as a kitchen. Each person has a particular strength like cooking or gardening that they focus on but are also responsible for contributing to all areas of the community, even if it's to a lesser degree. Some communities have people buy shares to become members; others work on arrangements such as working towards ownership or membership.

The concept of an intentional community is an excellent one. It's based on shared responsibility and using people's strengths to improve the community. It's part of the "two heads are better than one" train of thought, only this one can have multiple inputs. There can be challenges though, many of which the communities try to address right up front. They generally work on a democratic model, so that major decisions are voted on and majority rules. Human nature being what it is, personality conflicts can develop and egos can be bruised. If the community has been set up correctly and everyone knows the rules going in, this should help alleviate the problems. But people are people, so inevitably there are going to be challenges with certain personalities clashing or various visions of the community's direction coming into conflict with each other. Today, because so many of these arrangements are voluntary, members often come and go. This may very well be one of the best ways for humans to live in the future, in which case the advantages of someone living there may well outweigh the frustration with how it is governed. And there will probably always be gardens to be hoed by hand or firewood to be split for those who need an outlet for their aggression.

There are many in the prepping community who feel that having a number of strong and able-bodied individuals sharing one rural location is the best model to follow. This is based on a number of things. The first is that having a number of members or families pool their resources can allow for economies of scale in terms of many of the items a resilient homestead may require such as food supplies. Having more people to grow food in a worst-case scenario will be beneficial. If you were heating with wood and plentiful gas wasn't available for chainsaws the more people to cut firewood for warmth, the better. It's tough work when fossil fuels are available so doing it by hand is much more work.

One of the key elements of this concept though is protection. Should society cease to function the way we've become used to, there is likely to be an element of lawlessness. We will no longer be able to expect the same level of protection from law enforcement and potentially even the military. So having more people living in a homestead means more people to "man the ramparts" should a group arrive and decide they want what you've taken the time to grow or store. The people who may want to take your stuff may also decide it's easier to approach you at night under cover of darkness, so having more person power means you'll be able to have someone keep watch while others sleep and recharge for the next days' efforts.

This will sound very foreign to someone who has only seen these scenarios in movies about future dystopias (i.e. the opposite of utopia.) There are many, though, that feel that the veil of civility could end quickly with some large-scale economic or political event. They feel that living as part of a group will make you much more resilient to the chaos which may engulf your community. If you want to dip your toe in this water, you might want to read "Patriots: A Novel of Survival in the Coming Collapse" by James Wesley Rawles. While it is fiction it is an excellent exposure to the benefits of belonging to a group during "transitional" times.

Pooling resources and skills is a great idea. Many minds working together can work through more potential scenarios and outcomes. If you're thinking of organizing yourself in this way it's really important you have a clearly spelled out process for decision making and work division. My experience has been that many groups that have organized themselves on a "communal" basis eventually dissolve because of basic human emotions... resentment, feeling each member's input isn't respected, development of cliques, basically all the things that cause tension in society at large, just more focused in a smaller group. And remember these have happened during 'good times.' With the stress of a future scenario of scarcity and social breakdown, they will be amplified.

Renting vs Owning

For many of us, for most of our lives, owning a home was a safe secure investment. Some even got caught up in the frenzy of the housing boom in the run up to the 2008 economic collapse. For many the housing crash was a rude awakening. Now more than 4 years later some economists are suggesting that the market has hit the bottom and is starting to rebound. It can be difficult to determine the validity of such claims. If you don't own a home now though, you may still want to wait a little longer before jumping into the market. There is no indication that interest rates are poised to increase and there is still the possibility that housing prices may continue on their downward trend. It's hard to know whether banks which had foreclosed on homes may now start releasing them back on the market to try and recover some losses, which will keep house prices stable for a while yet.

While owning a home has historically been part of The American

Dream, 2008 turned homeownership into a nightmare for many people. What could be worse than owing more on your mortgage than what your home was now worth? That is devastating. So in a climate where mortgage rates remain at historic lows and the bottom of the market may not have been reached in your area, renting may still be a strategy worth considering. Save as much as you can in anticipation of getting back in when the bottom appears to have been established.

Conclusion

No one has a crystal ball to predict how the future will unfold. In the reference section of this book we've provided an extensive list of other reading material and websites that will allow you to begin expanding your knowledge of the challenges we'll face and how they may play out. You'll have to make your decision on your living arrangements based on your exploration of many differing views. From my biography earlier in the book you know that I've made the choice to live in the country, ten miles from a village and three miles from my nearest neighbor. This may seem a bit extreme to many, and it is partially because this was the off-grid house we could afford. I believe that the nearby village will be an excellent place to live in the future. It is the central meeting place for the many farmers who surround the village as well as for an eclectic variety of urban refugees who bring a varied and creative spirit to the village. Yes, we'll all be driving less, but there is land for the production of biofuels like biodiesel, and once in a while some locals ride their horses into town, not by necessity but by choice. So it won't be a stretch for many to return to some of the older ways.

Do your research, make your Plan, and start scanning those real estate listings for your dream home, wherever it may be.

12 Heating

Staying warm is a good thing.

It's a very basic thing. It's very low on Maslow's hierarchy of needs, but it's a big priority for many North Americans. It is going to become a big challenge for many of us as the sources of heat that we've relied on become less dependable.

Natural Gas

We are experiencing very low prices for natural gas right now, brought about by hydraulic fracturing or "fracking" of shale gas. Early in the new millennium the U.S. Energy Information Agency was suggesting that the U.S. had hit peak natural gas and that traditional production was starting to decline. Companies then began fracking, in which water, sand and chemicals are injected at high pressure deep into the ground to break up rock and release trapped shale gas. There is great concern about the safety of this process with the risk it poses to drinking water supplies and its potential to cause earthquakes. But many companies began aggressively fracking and were able to find enough gas to drive down the price of natural gas.

This type of deep and horizontally drilled well is very expensive so as the price of natural gas declines, the return on investment is decreasing. As a result some companies moved drilling equipment to look for oil, which remained close to $100/barrel for some time. These shale gas wells also have a very high decline rate, which means more money must be poured continuously into finding new wells. As oil plunged from over $100/barrel to close to $50 in 2015 some companies are finding it difficult to make money and had taken on great debt. These are challenging times in the fossil fuel industry.

So while we have a temporary glut of natural gas it appears that as the wells decline quickly and capital leaves the hunt for more shale gas, prices are destined to return to previous levels. The reality is that it's taking more money and more energy to drill wells and find what's left of the natural gas in the ground. The days of drilling easy, cheap, shallow wells and finding huge amounts of natural gas appear to be over, and over time heating with gas could be a much different ball game than it is today.

Home Heating Oil

How about home heating oil? Heating oil like natural gas is a carbon-based, non-renewable fossil fuel. As discussed in Part I, it appears that the world has hit peak oil and we will have to deal with ever-decreasing amounts of oil. Home heating oil is basically diesel fuel, which is produced when a barrel of crude is processed. Some of the barrel ends up as gasoline, some as other petroleum-based products such as jet fuel and asphalt, and some as diesel fuel or home heating oil.

So if you heat with oil, you'll be competing with trucks that use diesel to haul goods around the continent, municipalities that use trucks to plow snow and repair electricity networks, trains that burn diesel to move people and goods, and a whole host of other competitors that use oil, mostly in the transportation sector. North America is very transportation-dependent, from the movement of goods across the country to the fact that many people live in car-dependent suburbs that require driving to jobs and shopping.

So home heating oil does not offer any long-term security of supply. And price will be very volatile. Oil is an incredibly valuable resource and we were given a one-time endowment of it. For the last hundred years we have been pumping and burning the remains of plants that existed on the planet millions of years ago and through a miraculous process of heat and pressure gave us this incredibly powerful gift. But the endowment is starting to run out and there is nothing readily available to replace it. So there is no long-term future in heating with oil.

Now if you owned a restaurant or a french fry truck and had access to large quantities of vegetable oil from your deep fryer, then you would have an option, and that's biodiesel. That would require you to use "methanol," a natural-gas-based chemical that precipitates the glycerin out from the oil, leaving "fatty acid methyl esters" or biodiesel, which you could burn in your oil furnace. You would have to keep the oil inside so it didn't freeze in cold weather, but if you have access to large quantities of vegetable oil this is an option. You can learn more about biodiesel in William Kemp's *Biodiesel Basics and Beyond*.

Electricity

Some people heat their homes with electricity. This can take a variety of forms such as an electric furnace, baseboard heaters, or on-demand hot water heaters or boilers which heat the water that runs through radiators or in-floor radiant heating pipes.

Depending on where you live, this may be a cost-effective way to heat your home. At least it may be less expensive than fossil fuels like gas and oil. But electricity can be expensive to make, so its price, while reasonable today, is by no means guaranteed.

Electrical utilities have not been making sufficient investment in new generation or maintaining their existing generating facilities. So most North American electrical utilities will have to make massive investments in the future to keep the lights on. They have also not been investing in the infrastructure of transmission lines that bring the electricity to your house.

After the blackout on August 14, 2003, which left 50 million North Americans without electricity, President Bush called for spending of up to $100 billion to fix the system. But the money that would have funded a reliable power grid was spent on the wars in Iraq and Afghanistan.

So the grid is going to become strained as demand for electricity grows and the infrastructure to deliver it has not been upgraded. Plants that supply the bulk of our electricity are getting older and will become less reliable and more costly to maintain. And someone who cares about their footprint on the planet should not be heating with electricity when so much of it is produced by coal.

In the United States, 50% of electricity is generated with coal. This is a very dirty way to generate electricity, and even though you don't produce any CO_2 at your house when you turn on your electric range to cook dinner, when you use coal-produced electricity you are having an extremely negative effect on the health of the planet.

Those in districts that have nuclear power have another conundrum. Yes, nuclear power doesn't generate any significant greenhouse gases but it does create a legacy of nuclear waste. No government in the world has been able to devise a plan to safely store spent nuclear fuel during the thousands of years it will remain lethal. When your electricity comes from nuclear power, you contribute to this nuclear waste, which is also the material used to make nuclear weapons. There are lots of downsides to the nuclear option for power generation.

So the traditional heating sources—natural gas, home heating oil, and electricity—don't have a good future as potential heat sources for your home. Remember that gas, oil and the coal used for power generation are non-renewable fossil fuels. When you burn them you extract a carbon-based fuel that has been trapped beneath the surface of the planet, not doing any harm to the atmosphere, and you release $C0_2$. Carbon dioxide is the leading cause of global climate change. Even if you burn natural gas, which is considered "cleaner" than many other fuels, the reality is you are still contributing to climate change in a fairly big way.

Wouldn't it be nice if you could find a way to heat that offered more security and didn't do as much damage to the planet? There are two readily available forms of heating that use renewable energy: geothermal and wood.

Geothermal

When the sun hits the earth, the ground absorbs some of its heat energy. If you were to dig down five or six feet, you'd find the earth stays a fairly consistent 40° F. (5° C.) to 50° F. (10° C.) year round, regardless of the conditions six feet above. If you circulate liquid through a series of pipes in the ground, you can use that heat for your home and, best of all, run the system in reverse and use it to cool your house in the summer. This is an unlimited energy source; you merely have to use some energy to bring that stored solar power into your home.

With a geothermal system, or "heat pump" as some people call them, you pay roughly 25% of your home's heating and cooling costs while the rest comes free from your backyard. Here's an example: for every 1W of electricity you purchase from the electricity grid (to power the pump that moves the liquid) you get 3W of heat energy from the ground, giving you a total of 4W of heating or cooling for your home. Yes, you do still require some electricity, but this is much more efficient than using only electricity in resistive heating like baseboard heaters.

Geothermal systems require a loop to run fluid through the ground to extract heat. These can be horizontal loops, vertical loops, or pond or lake loops. In a horizontal loop, trenches are dug six feet deep and pipe is laid in loop circuits. In a vertical loop a series of holes are drilled in the ground and a pipe is placed vertically in the boreholes. In a pond or lake loop, the pipe is placed in the lake or pond and anchored offshore where it can't freeze.

While it may seem counterintuitive that the cold water of a lake in the winter could heat your house, the trick with a geothermal system is that it uses a compressor to operate a refrigeration cycle that extracts heat energy from the ground or lake and upgrades its temperature to a level high enough for space heating. In the summer, the process is reversed to provide air conditioning. The unit may also provide a portion of your domestic hot water.

The cost of a geothermal system may be more than that of a traditional fossil-fuel-based heating system like a natural-gas forced-air furnace, but this is not how to evaluate the purchase. That furnace will require you to continue purchasing a nonrenewable fossil fuel that most indicators suggest will increase in cost dramatically in the future. You should not use the purchase price but the lifecycle cost of the heating system, what it's going to cost to operate over its lifetime. Also remember that with a geothermal system you are purchasing both heating and cooling. If you compare the cost with that of a new high-efficiency furnace and a central air conditioning system, the geothermal system will be very competitive without the long-term commitment to the continued purchase of fossil fuel.

Wood

There are lots of misconceptions about wood heat. Some people think it is very polluting. Others think you can't heat a house with wood because they've experienced a decorative fireplace in the living room that actually sucks heat up the chimney rather than delivering it to the house. But a modern EPA-certified wood stove will burn very cleanly and, if situated correctly in your house, do an excellent job of heating it.

In the 1970s, during the first energy crisis, there was a renewed interest in wood burning, and the market was flooded with a variety of wood stoves. While many looked great and did a reasonable job of heating, the technology has continued to improve and evolve to the point where wood stoves are an excellent alternative to other heating systems.

Wood stoves are one of the most environmentally sustainable ways to heat because they use a renewable fuel—wood. When a tree is growing it uses photosynthesis to convert carbon dioxide from the air and store that carbon, while releasing oxygen. When the tree dies it releases the carbon it stored while growing, making it "carbon neutral." If a tree dies in the forest and falls to the forest floor, it will release the same amount of carbon that it absorbed during its life, and it will release the same amount of heat energy as it would if you burned it in a in proper EPA-certified heating appliance. You're just speeding up the release of that heat when you burn it.

The two main types of wood stoves are catalytic and noncatalytic. A catalytic wood stove uses a catalytic combustor similar to the catalytic converter that helps to decrease the pollutants produced by your vehicle. In this type of wood stove the wood is burned in a primary combustion chamber and the smoke is then diverted through the combustor, which is usually a honeycomb shape and made from refractory material that gets very hot. As the smoke passes through the honeycomb, any unburned gases or particulate gets burned off, making the remaining exhaust very clean.

In a noncatalytic wood stove there is just one burn chamber, but oxygen is drawn into it so that after the primary combustion there is a secondary combustion to clean up the exhaust before it goes up the chimney. I have had both types of wood stoves and much prefer the noncatalytic. Our catalytic wood stove required us to put it into an airtight mode after it got hot enough. This made the stove more labour intensive. It also required us to purchase the catalytic combustor on a regular basis for about $200. We were told that combustors should last close to five years but we rarely got more than two years of service out of one. The other downside to our catalytic stove was how many places air could get in. There were numerous gaskets that had to be replaced to prevent outside air from getting into the stove and affecting its performance. If too much air gets into the stove it burns the wood faster and therefore requires loading more often and goes through more wood. This stove was time consuming and expensive to maintain and never worked as advertised.

We now use a Pacific Energy wood stove and it is exceptional. It has only one gasket to be maintained, which is on the front and only door. It burns exceptionally cleanly and when loaded with hardwood burns for 10 to 12 hours. (We could rarely get more than 6 hours from our catalytic woodstove.) Now we can load it up at night, have a sound sleep, and get up to a bed of glowing coals. We just toss in some more wood and have it roaring again in minutes. It's quite brilliant.

People probably assume that heating with wood is an option only if you live in the country. There are companies that will deliver firewood to cities, so it may be worth tracking some down and getting a handle on costs. There is also a huge untapped source of firewood in urban areas that comes from people trimming and cutting down trees on their property. This is often left on the lawn for someone to grab. A simple knock on the door when you see the tree being cut or notice a pile of unclaimed firewood often gets a response of "That would be great; I'd be so glad to see someone haul it away."

There are lots of trailers you can buy to haul this wood home. And cities have made sourcing this wood even easier by having pickups of "organic" materials periodically. This is often in the spring when people are trimming things back. I'm amazed at the size of some of the wood I see people bundling up to truck away. It's yours for the taking and it offers a huge untapped heating source for your home.

You may not have a huge choice about the quality of the wood, but if it's free, who cares? Ideally you want hardwoods like maple and oak, which will burn longer and have more heat potential. Softwoods like poplar or birch won't produce as long a burn, so you'll have to burn more of them. If you have both in your woodshed, you'll like the softwoods for the fall and spring when it's not brutally cold and hardwoods for the coldest months of winter. If all you have access to is lower-grade softwoods you'll just have to budget more space for storage since you'll need more. Many areas that don't have as much hardwood have more temperate climates, which helps.

If you are harvesting your own firewood, the size of woodlot you need depends on a number of things: how big your house is, how well insulated it is, whether you like to walk around in a t-shirt in January or a nice thick sweater and wool socks. It also depends on what type of wood you have on your property and how well managed it is.

A fairly typical North American house needs around 20 acres to produce a sustainable woodlot. This assumes a good mix of trees and selective harvesting, meaning that you cut larger more mature trees and allow the light to get to the smaller and lower trees so that they grow more vigorously to fill in the gaps. To properly thin the woodlot you should be looking for poorly formed, crooked, insect-damaged, dead, and undesirable trees. While you don't need to aim for the perfection of a suburban lawn, removing trees that don't have the long-term potential to be large and vigorous trees makes for a healthier woodlot that can provide an endless supply of heat for your home.

Sound woodlot management also assumes you are not harvesting with a huge machine which kills all that lower growth. You can hitch a horse up to lengths of trees and haul them out of the woods and up to the house.

I use a large plastic sled, which was designed to be towed behind a snowmobile, but is very light and rugged. Once the tree is down I cut it into longer lengths which will yield two or three fire-sized logs. I pull this out to the house or to where I can get to it with the truck in the spring. I'm careful not to disturb too many of the smaller trees and seedlings on the ground. Once I get these longer lengths back to the house I cut them with an electric chain saw. There is nothing more satisfying than cutting a carbon-neutral heating source like firewood with an electric chain saw powered by my solar panels.

Pellet Stoves

Some people have discovered heating with wood pellets in a pellet stove and it's an excellent option as well. Pellets are made by compressing wood waste from sawmills into smaller pellets that look like rabbit food. They are fed into the burn chamber of the pellet stove by an auger. These stoves require electricity to power the auger that delivers pellets to the firebox and the fan that blows air out of the stove, so make sure you have a backup source of electricity to keep your home warm during a power disruption.

Pellet stoves and pellet sales seem to be very irrational. Every time there is a spike in natural gas prices it is hard to buy a pellet stove because they are snapped up. Then the next year if natural gas prices settle down pellet-stove sellers will have a warehouse full of unsold stoves. The pellets themselves are also prone to the vagaries of the market. When they are abundant, the price is very affordable. When they are in short supply the price rises. Much of the wood waste that they are made from comes from the mills that make lumber for new homes. With the downturn in the North American housing industry these mills are not creating as much waste, so the supply of pellets has been severely restricted in many parts of North America. People who bought pellet stoves thinking they'd removed themselves from the ups and downs of the natural gas market are coming to realize there really is no shelter from the fossil-fuel-depletion storm.

As the price of all energy rises, consumers will "fuel switch" and they'll do it often and very creatively. If natural gas gets too expensive, they'll buy pellet stoves. If wood pellets get too expensive, they'll plug in space heaters and use more electricity. The reality is that it will be difficult to make a long-term plan for heating your home unless you have access to your own woodlot or heat with a ground-source heat pump and have your own significant renewable energy system to generate the electricity you'll need to move the liquids through the heat pump to recover the heat. Heating with oil, natural gas, electricity, or pellets or firewood that you purchase from someone else will become a wild ride in the future, and the direction of the prices you'll be riding will be constantly upwards.

Time for a heating "Plan B"

If you currently live in an urban area where you heat with natural gas, it's a good idea to have a "Plan B" heating source. Perhaps it means making sure you have some space heaters to keep rooms you spend most of your time in comfortable. And if you believe as I do that the electricity grid will become less dependable after years of underinvestment in new generation and infrastructure, it wouldn't hurt to have a "Plan C." That might be a wood stove in the basement for those emergencies. Every time you take a trip to the cottage bring half a trunk of firewood home, and when that neighbor down the street finally decides to take down the silver maple that keeps dropping branches on his new car, make sure you're there with your wheelbarrow to haul away the fruits of his labor. It's free, and if you burn it in an EPA-certified wood stove, it's carbon neutral and an excellent way to heat.

Heating your home in the future is going to be a challenge. It's going to require more of your time to think about how you're going to do it, and it's going to require more of your time and income to accomplish it. Cutting your own firewood will be cost effective, but it will require more of the time you formerly used to earn an income, or watch football games on Sunday. If you now cancel your health club membership because you're getting such a great workout cutting, hauling, and splitting firewood, all the better.

What's important is that you realize that we are coming into a time of instability and increasing prices for home heating. Analyze where you live and what resources are near at hand. Do you live near lots of forests? Are you close to a source of natural gas or does it travel a long distance? Has your local utility been investing in new electricity generation so that a reliance on it as your primary heating source may be realistic? How reliable does the grid seem where you are? Pick what looks to be most cost effective over the long haul, and then spend more money on installing your "Plan B" backup heating source. And if you're like me, make your family think you're crazy and have that "Plan C" system ready to go. There's nothing worse than being cold, and with the knowledge you now have about how uncertain the future may be, you have no excuse to ever be cold.

13 Fuel/Energy in Your Home

For some people reading this book, heating will be less of a concern than cooling their homes. According to the US Department of Energy the average North American home uses 52% of its onsite energy in heating, 22% for appliances, 17% for hot water, 4% for refrigeration, and 4% for cooling. If you live in Arizona obviously your cooling requirements will be much higher, but this is an average.

While some of the energy requirements other than heating, like hot water, will be supplied by natural gas, many will be provided by electricity, which, like all large-scale energy sources is going to be challenged in the future. Some of this may be fuel related and some of it may just be that operating things on a large scale is going to be more difficult in the future. Many utilities have not been investing properly in their infrastructure. The transmission lines, transformers, and all the equipment needed to get electricity from the power plant to your home is like anything mechanical and it needs maintenance. And as demand has grown the infrastructure required to move higher volumes of electricity hasn't been upgraded fast enough to keep up, so the system is taxed.

As storm intensity has increased with a warming climate, power outages are becoming a more regular event for many people. So you should start assuming that electricity is going to be more expensive—a reason to use less—and that it's going to be less reliable. This should be all the incentive you need to start making yourself more independent when it comes to electricity. Lots of people are already inspired to put up solar panels to do the right thing for the planet, but there are important steps to take before you put up solar panels. Using the guide created by William Kemp in his book *The Renewable Energy Handbook* you need to follow the "eco-nomic" approach to adopting renewable energy. And that means starting with "nega-watts" or energy saving first, which you accomplish through energy efficiency. This is the "sensible" approach to

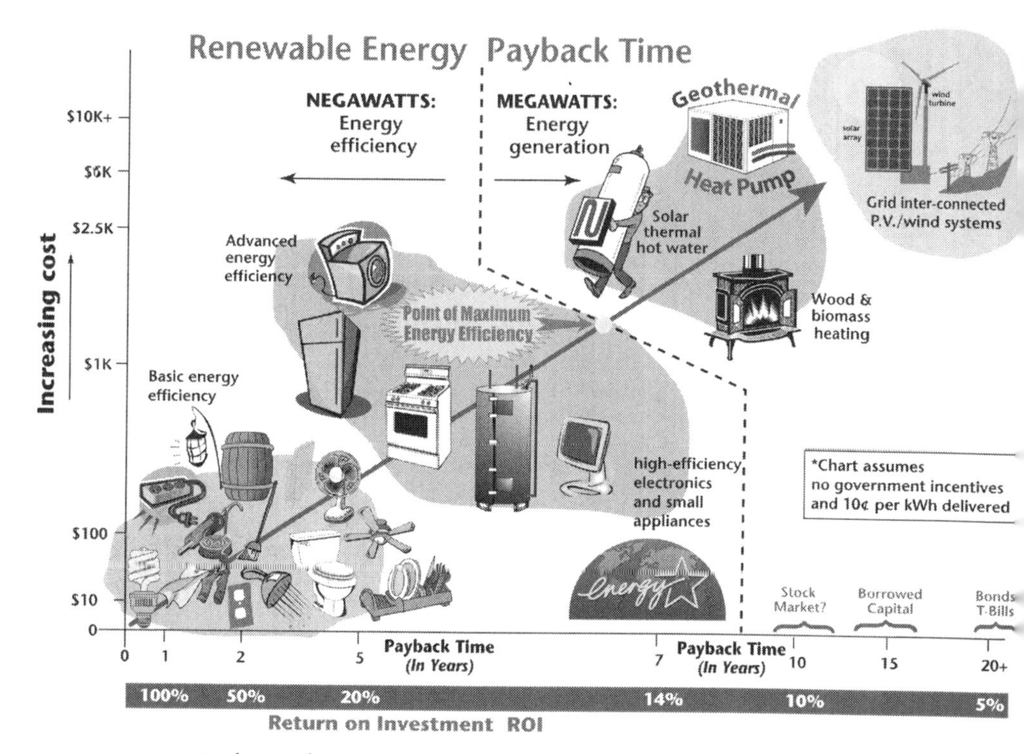

energy independence.

Most people want to start putting up solar panels to generate electricity, but if you follow this chart you'll see that you should really start with simple, inexpensive solutions to reduce your energy consumption first. It's less sexy and not as easy to brag to your neighbors about, but saving energy has a much faster payback than generating energy. That's why one section of the chart is labeled "negawatts," a term coined by Amory Lovins and The Rocky Mountain Institute to represent saved energy.

As you think about those items that give you the fastest payback, a light bulb should appear above your head. It should be a compact fluorescent light bulb (CFL). The traditional incandescent light bulbs that use a hundred-year-old technology are much better producers of heat than light. You can tell this by examining your kid's Easy-Bake Oven. It uses a 100W light bulb to bake cakes! And we've continued to use these bulbs as central air conditioning has become more common, so you have this double whammy for your house. You're using electricity to cool your house, then you're using more electricity, very inefficiently, to light it, and then you have to use more electricity to remove that extra heat you've created with incandescent bulbs. Countries like Australia and Canada have begun a phase out of these older incandescent bulbs with an eventual move to a complete ban, so the writing is on the wall.

Compact fluorescent light bulbs (CFL) are a great example of efficiency, being five times more efficient than incandescent bulbs and lasting ten times longer. So even though the initial purchase price is higher than the price of incandescents, over the life of the bulb you'll save a lot of money. LED light bulbs, which are becoming more readily available, are even more efficient than CFLs but since they are still generally more expensive compact fluorescents are more economical.

The August 2005 issue of *National Geographic* magazine included a photograph of a man in front a huge pile of coal holding a compact fluorescent light bulb. The caption read, "A CF light bulb lasts 10 times longer and saves nearly a quarter ton of coal over its lifetime." I contend that everything I'm recommending in this book has three huge benefits. Replacing your light bulbs with CFLs is going to save you money. This helps you deal with the challenges we're facing economically. It also significantly reduces your carbon footprint. That's good for the planet. In this case it also helps you reduce the costs to operate your home. Good for your pocketbook. Good for the planet. Good for reducing your operating costs to help make you more independent.

As you read the energy-efficiency chapters of a book like *The Renewable Energy Handbook* you discover numerous ways to increase your home's efficiency and save money. Some of these involve water-saving devices like low-flush toilets and low-flow showerheads. In the city the cost of water is generally so cheap that you will not notice a huge difference in your bill if you reduce your water consumption, but that will change as municipalities find it increasingly expensive to maintain aging infrastructure. In the country, water can be a much bigger issue. In our off-the-electricity grid home one of our largest uses of electricity is our deep-well pump, so using water efficiently is really important.

Basic energy efficiency also allows you to start evaluating just how committed you are to a path of independence. There are a number of products in your home that have a negative impact not only on the planet but also on your ability to be independent. They are machines that you have to keep pumping money into even though there are perfectly good alternatives that only cost you your time. Many see something like using a clothesline rather than a dryer as a step back in human progress but in reality many of us are going to have to relearn these skills anyway. The question is do you want to embrace it yourself or have it forced upon you?

Now a clothesline, there's an elegant piece of technology. A simple way to accomplish a task naturally, letting the sun and wind do what they do naturally without you having to purchase fossil fuels. And you don't even need the sun and wind for that matter. There are numerous drying racks that allow you to dry clothes in your house or apartment. You just drape the clothes over the rack and moisture escapes into the air. When you use these in the winter it's a nice way to add to the air moisture that your furnace or woodstove removes.

We don't need dryers, and we don't need dishwashers. They are a carbon intensive luxury that as you scale back and stop using allows you to be much less affected by interruptions of the energy that keeps them working.

Your home is actually full of appliances that do things you could do for much less money and with a much smaller carbon footprint. Ceiling fans use significantly less energy than an air conditioner. Using a push mower to cut your grass has a huge impact on reducing pollutants and $C0_2$ emissions. A rake will move leaves with much less damage to the planet than a leaf blower. Same with a shovel for snow removal rather than a snow blower. Start looking around your home and figuring out what you can do manually rather than using some machine. And you'll not only protect the health of your pocket book, you'll also protect your own personal health and the health of the planet. All the activities that we're talking about are much better for you. Sucking in the exhaust from a lawn mower or a leaf blower's internal combustion engine is not good for you.

Sometimes the simplest solutions are the best. Good for your health. Good for the health of the planet. Good for your pocketbook.

Another way to look at all these actions you may take to make your home more energy efficient, is that it's much easier to power a home that uses fewer energy so it's easier to do independently. Using less energy means needing fewer solar panels. And for those times when the power grid isn't there and you're powering your home with a generator, you'll be able to keep more things working if they all use less energy.

Once you've worked through the basic energy-efficiency savings it's time to look at advanced energy efficiency, which includes appliances. Appliances have a natural life and you have to replace them eventually. Tough economic times will force many people to hold on to them much longer than they otherwise would, but if your fridge is a beautiful 1970s avocado green, it's time to replace it. Over the long haul as the cost of electricity rises older models are going to be huge drains on your pocketbook.

When it comes time to replace your appliances remember that it's not only the purchase price of a product you have to consider, it's the lifecycle cost of that product. What will it cost to operate over its lifetime? If you buy the most efficient washing machine, with the money you'll save in operating that washer over its lifetime, you could buy three of the least efficient machines. This assumes you are using hot water and have four people in the home. You'll also get a fivefold reduction in water use and a tenfold reduction in soap use.

The way to determine if the appliance is the most energy efficient is to use the EnerGuide rating that should be posted on or in the appliance. This will show you how this particular model stacks up against other comparable models in its class. You want to make sure you buy the one that uses the least energy. Sometimes you will not be able to buy the very best, but as long as you're close you're on the right path to keeping your ongoing energy costs down.

You probably won't be able to replace all of your appliances right away, but as they get towards the end of their life it's important to replace them with the most efficient models. You should also be looking for the EnerStar logo as an indicator of efficiency in the product class. This is helpful when you're looking at electronics like TVs and computers. Laptop computers can be up to five times more efficient than desktop computers. If you have to use a desktop make sure you look for an LCD screen similar to the kind laptops use rather than the traditional large CRT type.

All these things add up to big energy savings in the long run and to you keeping more of your after-tax income. So why not quantify that savings to reward yourself for your accomplishments and inspire yourself to move onto the next step? I recommend that you open a brand new bank account and call it your "Green Energy Account" or your "Solar Powered Bank Account." Put all the money you've saved through your energy-efficiency activities towards products that will now generate green power and make you even more energy independent.

So before you start any of your energy-efficiency projects, make sure you keep the last few years' worth of energy bills. Then, as you begin reducing your electricity consumption, calculate how much money you are saving month to month. This can even extend beyond electricity consumption. If you have had an energy audit done on your home and you have upgraded insulation or improved weather stripping and replaced leaky windows, try to compare with previous years energy consumption and calculate the savings. This is money that you've earned and paid tax on, so these savings represent real progress in your move towards independence on all fronts.

As you go through the process of making your home more energy efficient, eventually you'll hit the point of "Maximum Energy Efficiency" where you've pretty much done all you can to save every kilowatt of electricity and every penny of wasted income. This has probably taken a year or two, especially with appliances, but you should now have accumulated a nice amount in your Green Energy Account. So if you go back to the chart on page 106, you'll see that you're now ready to start generating some of your own energy rather than just saving it.

Solar Domestic Hot Water

Most people assume that putting a solar panel on your roof is for generating electricity, but the first panel you put on your roof should actually be to make hot water. This is called "solar thermal" or solar domestic hot water (SDHW), meaning that it is for your daily hot water needs like washing clothes and dishes and bathing, not for heating your home. If you've ever picked up a hose that's been on the ground on a sunny day and felt how hot the water is when you first turn on the tap, you know about the sun's ability to heat water. Using this principal you will actually get a much faster payback when you invest your green energy dollar in a solar thermal system first.

The system consists of a collector that goes on the roof to capture the sun's thermal or heat energy. Then there's a system to transfer that heat into your hot water tank. Some systems pump the water through tanks with coils inside to heat the water; some use heat exchangers on the outside of the tank. There is also a choice of two main systems to go on your roof. One is a flat-panel collector and the other consists of vacuum tubes. Both have advantages and disadvantages, and some research about your local conditions and a discussion with local dealers will give you an idea of what is right for your home. The main systems on the market now are all 12-month units, meaning that they work year-round. These systems circulate a food-grade antifreeze called propylene glycol through the system and then transfer the heat to your domestic hot water supply.

Early solar thermal systems just ran the water for your home through a panel on the roof to heat it. Obviously it would not have been a good thing to have water freeze in pipes on your roof if the temperature got below freezing. So these systems were often referred to as "drainback," meaning that once the sun went down or on cold and cloudy days the water would drain back from where it might freeze and then be pumped back into the collector once it was safe.

Solar thermal systems are simply amazing products. There is nothing more gratifying than trying to hold onto a hot water pipe that is bringing sun-heated water into your house, with no fossil fuels burned and therefore no greenhouse gases flowing out of your chimney or greenbacks flowing out of your wallet. And best of all it increases your independence from rapidly rising energy costs and potentially disruptive shortages of the fuel source you use for hot water. The size of the system you should install will depend on the number of people living in your house and how extensively you use hot water. A house with four teenagers will use a lot more hot water than a home for an older couple. To a certain extent, some people may find that they start using hot water in tune with the weather. If you wait until a sunny day to wash your clothes, you'll have lots of hot water as well as heat from the sun to dry the laundry on the clothesline. This is a good thing.

Some people may even attempt to build their own system using off-the-shelf components. These units are usually less sophisticated and are drainback systems to avoid having to charge the system with propylene glycol, but that's not to say that if you're up for the project you shouldn't do it. If you've lost your job and are having trouble finding work but your spouse is employed, you'll be no doubt cutting back on expenses everywhere. This may be a case where you invest your time and some money which over the long haul offer you an excellent payback. You're just taking the time you could otherwise spend earning an income to pay someone else to install your system and using it instead to learn how to do the installation, allowing extra time to make some mistakes along the way.

Standing under a shower where hot water is being heated by the sun is a totally awesome thing. You are not releasing any emissions of CO_2 into the atmosphere, and the shower isn't costing you a penny. Investing in a solar thermal system is one of the soundest financial commitments you'll make. Your retirement fund statement showing you how much your stocks have gone down may make you red in the face, but it won't be as comforting as a hot shower on a cold, sunny day.

On Bill Kemp's chart on page 124 you'll also notice that wood stoves and ground-source heat pumps also appear in the "thermal" section. Both of these make use of solar energy and they are discussed in more detail in the heating section in Chapter 11.

Solar & Wind Electric

The second type of "ECO-nomic" payback afforded by renewable energy comes from using the sun, wind and water to produce electricity. While you may invest anywhere from $3,000 to $7,000 in a solar thermal system, a figure from $10,000 or $30,000 is more in line with what many invest in solar system for electricity. There is still a payback, but it takes a little longer. The comfort and security this will bring to your home is priceless.

Solar or photovoltaic (PV) panels are the best place for most people to start using renewable energy to generate electricity. A solar panel converts the sun's energy into DC (direct current) electricity, but most of the appliances in your home require AC (alternating current). You use an inverter to convert the DC to AC so that you can power all the AC appliances in your home. In the early stages of the industry some people used the DC electricity directly and purchased DC appliances, which were commonly used in the recreational vehicle industry at the time. Early inverters were quite crude and inefficient, so early on this made sense, but modern inverters are extremely efficient and there is minimal loss involved in making the conversion, so it only makes sense to go AC.

Many levels of government now provide financial incentives for the installation of renewable energy, so you need to research what is available where you live before you proceed. Some incentives provide a rebate on a portion of the purchase price of equipment, while others actually purchase your "green" power for a specified price over a given contract length. Your dealer will be able to help you calculate the savings this will represent. The Database of State Incentives for Renewables and Efficiency (www.dsireusa.org) is an excellent resource to find incentives in your state. In Canada the Office of Energy Efficiency of Natural Resources Canada provides a great listing of incentives (www.oee.nrcan.gc.ca.)

As outlined in *The Renewable Energy Handbook*, you'll need to calculate your electrical loads before a dealer can properly size your system. This is where taking all that time to reduce your electrical usage will pay off. For every dollar you spend on energy efficiency, you'll have to spend about $5 for generation. In other words, you can continue to use inefficient incandescent light bulbs, but you'll need to spend $5 more to purchase solar panels versus the $1 you spend on energy efficiency to reduce your demand.

Once you've reduced your electrical requirements you can make your investment in solar panels. It is an investment. In fact with the rapidly declining price for PV panels this is an excellent time to invest. And best of all, they will be providing a huge degree of energy security for your family. Anyone who has experienced a blackout knows how quickly the inconvenience of being without power turns into a huge problem. Solar panels are hard assets, and hard assets have numerous advantages. When it comes time to size your system I recommend you oversize it. Having more power than you need is not a problem and you can always find productive things to do with it, whether it's pumping it back into the grid or preheating hot water.

You should not wait to take the plunge. Solar panels work! They are dramatically cheaper than they were just a few years ago. They make electricity… today! They help inflation-proof your family budget, today. They reduce your carbon footprint today. They'll power your home to keep the lights on next time an ice storm takes down power lines or a hurricane causes outages. They are an indispensable tool for your home, today. They are the ultimate "hard asset" for challenging times.

Along with your solar panels you'll need a few other pieces of equipment like a charge controller and a DC disconnect to isolate the solar panels from the rest of the electrical system. While it's possible to not have batteries connected to your solar panels, I would recommend you have some. If you're connected to the grid you won't need as many as if you're off grid, but you should have some nonetheless. Someone who lives off the grid will need a larger capacity for their batteries to allow them to run their household for several days of cloudy weather. They might spend $4,000 or $5,000 on their battery bank. Someone who is connected to the grid should use the grid as their large battery storage, putting excess electricity into it during the day and taking out what they need at night. They should also have a small battery bank, which would cost under $2,000, to allow their household to run should the grid go down but not have more storage capacity than they need.

If you're grid-connected you should also get an electrician in to rewire your panel box to have "essential" and "non-essential" loads. This will ensure that your batteries will only have to power loads that are absolutely critical for you during a power outage. These would include your fridge and freezer, a furnace fan if you heat with a forced-air furnace, lights, and perhaps a TV or radio so you can stay up-to-date on what is happening to get the power back on. Sizing the battery bank to power these loads will depend on how efficient they are, so you should be powering a very efficient refrigerator and light bulbs to make sure you aren't wasting any of the limited power you've got in your batteries. It will also depend on what type of outages you expect. If you're used to summer storms leaving you without electricity for a few hours it will be quite a bit different than if you experience hurricanes or ice storms leaving you in the dark for days.

The nice thing about having essential and non-essential panel boxes is that it will make it easier if you decide to add a generator to the mix which I discuss in the next chapter.

Another excellent way to keep those backup batteries charged and help with your energy independence is a wind turbine. I should qualify this by saying that wind is excellent if you are in the right location. It's the dream of many people to have a wind turbine at their home, but if you live in an urban area this may not be realistic. The best wind is the wind that has had a long open area to travel over before it hits a turbine. A large body of open water is the best location. If you live beside a large lake or river you are probably in a good spot for wind. If you live in an area with lots of open fields wind is something to consider. Wind also accelerates up a hill, so if your home is perched on top of a hill consider an investment in wind. What wind doesn't like is obstructions like forests and barns and houses. Wind flows like water in a stream. When water encounters obstructions like rocks it slows down. The rocks create turbulence and reduce the flow rate, known as "laminar flow." Wind operates the same way; every time it encounters an obstruction, turbulence is created, which interferes with its flow.

Urban areas are composed of endless obstructions that create turbulence, so they are generally not a good area for wind. Many cities are located on large bodies of water, but the areas around the water tend to be the most economically desirable, so putting wind turbines there poses a problem. If you work in a large building downtown you know how windy it can be around those office towers, but this is not good wind for power production because it is so erratic. All that turbulence would confuse a wind turbine that wants to have a strong consistent wind turning its blades from one direction.

Most wind turbines today are the well-known three-blade horizontal-axis turbines like the very large ones you see on the wind farms that are becoming more common in many parts of North America. They represent the end result of years of experimentation and refinement. Many designs have been attempted, but large turbine manufacturers have settled on this design because it has proven to be the most effective way to generate electricity from the wind. Most smaller home-sized wind turbines follow this design as well.

As is so often the case when you have a sudden interest in something like the environment, the market is flooded with products that seem to help solve the problem but have unproven technology. If you are looking to purchase a wind turbine for your home you'll want to see its "power curve," which will show you how much power it's capable of producing over increasing wind speeds. Most turbines will ramp up to a peak output and then drop off as the turbine reaches its maximum-rated speed, stalling or furling to prevent it from damaging itself in excessive winds. Some of the newer vertical-axis turbines have been slow to produce power curve data, which means that you should be skeptical about their rated performance. The home-turbine market is dominated by several manufacturers that have been producing smaller-scale turbines for many years, including Bergey Wind, Southwest Windpower, Kestrel, and Africa Wind Power. If you are considering a wind turbine I would recommend that you stick with one of the main manufacturers with a proven record. These companies all produce turbines that use the standard horizontal-axis design.

The correct way to determine if your site has wind resources that make the investment in a wind turbine worthwhile is to put up an anemometer at the height you anticipate installing the turbine and logging the wind speed over time. This can be cumbersome, and often people go to online resources that include wind maps of your area. Various levels of government have maps which allow you to decide how good your local wind resources are. Many people say "the wind always blows at my house," but whether or not that wind has good potential energy in it remains to be seen. A wind turbine should be 30 feet above and 300 feet away from the nearest obstruction such as a barn, silo or a forest to ensure a good clean flow of air and to maximize the potential of your turbine investment.

Several years ago we put up a new Bergey XL1 (1 kilowatt) wind turbine. We had a smaller turbine on a 60-foot tower that never lived up to its potential. It broke twice, and the second time I decided it was time to go big or go home. We installed the Bergey on a 100-foot tilt-up tower. We are not in a prime location for wind. In fact I believe we would be considered a poor area. We are 45 minutes north of the Great Lakes, which are an excellent wind source, but we are too far away to take advantage of them. Forests, thousands of acres of forests, also surround us. While these make for an amazing place to live, they are not conducive to wind power. Our tree line is about 60 feet, so we felt that with the tower 30 or 40 feet above them we at least had a shot at some wind.

Living off the electricity grid we also have a unique perspective on wind. In any off-grid home having a hybrid system that utilizes both solar power and wind power makes sense. The months that have the most wind, in our case the fall and winter, also have the least amount of sun and vice versa. Before installing the new turbine we needed to run our gasoline generator about 15 times during the fall months in order to charge up our batteries during periods of cloudy weather. After installing the turbine in September 2007 we ran the generator only three times, and then went ten months without running it at all. This was quite an accomplishment for us. Our desire to reduce our generator run time comes not only from a desire to add less carbon to the atmosphere but also from a fear of where we see the price and availability of oil going. With most oil producing countries past peak and many in steep decline, I believe I eventually won't be able to afford to run the generator, if I can find gas at all. So our wind and solar ensure that our lights stay on, our pump keeps water flowing in the house, and the fridge and freezer preserve our food.

Installing the wind turbine was a huge challenge because although I am not an engineer I decided to install the turbine myself anyway. Many local dealers are at the stage where they don't want to be bothered with wind. Wind turbines are mechanical instruments, which means they break. And when they break they are usually located at the top of poles and towers that make them difficult to reach. With solar panels, on the other hand, once you mount them on a roof or tracker they just work. This might be something you should bear in mind as you evaluate various technologies to make your home more independent.

After living off the grid for ten years I had developed a certain amount of confidence that I was up to the task and knew it would save a lot of money if I installed the turbine myself. With our video publishing business, we also decided it would make an excellent opportunity to produce a new DVD called *Installing a Home-Scale Wind Turbine*, which we added to our catalog. While putting up a wind turbine is a lot of work, videotaping the whole process doubles (or triples) the work because you have to move the equipment into position, bring out the tripod, set up the camera, set up the shot, and half the time do the step you're working on more than once to a get a good take.

We had to use a backhoe early in the process to level out the area where the tower was going since it helps if all four anchors are roughly level. Then we had to dig six holes, including one for the base and one for the winch. I elected to dig these holes by hand because we were experiencing a drought and our sandy soil was collapsing into the hole when we had the backhoe attempt it. I learned that digging a 5-foot hole to get below the frost line is easy in sand for about the first 2½ feet, but when you're only 5'8" you can start getting tired rotator cuffs pretty quickly when tossing sand over your head. We built rebar cages to strengthen the concrete we poured into the holes.

Once the anchors were in I had to dig the trench from the base of the tower to the battery room, and at this point I decided it was time to get the backhoe back. Then the tower had to be laid out and guy wires prepared for the lift. We lifted our turbine with a permanently installed winch which we power with an 18V Dewalt cordless drill. We can raise the tower with four batteries and lower it with two, so it's an exceptionally efficient winch. The nice thing about our gin-pole-type tower is that we can raise and lower it fairly conveniently. This helps if you ever have to do maintenance on the turbine or if you are in a hurricane-prone area. You could have the tower down in about half an hour if a huge windstorm was approaching.

When the tower was finally up I must say it was one of the most gratifying things I had ever done. For years we had been exhibiting our books at renewable-energy fairs and usually took solar panels and our old wind turbine there to attract attention to our booth. This is where I learned the male/female side of renewable energy. If you have a working wind turbine at a show, men will be drawn to it like moths to a light bulb. They will spin it as hard as they can, stick their hands into it to see how much it hurts when they try and stop it, and try and figure out how to make their own turbine from the broken car alternator in their garage. Wind turbines are very male. They can be loud, make a lot of commotion, and scream "look at me!" Women, on the other hand, are drawn to solar panels, which work away quietly, generating electricity calmly and efficiently. They just do the job. They don't make a lot of noise, don't draw attention to themselves, just work away quietly in the background. With two daughters, I have always considered myself a feminist and this certainly held true with my solar panels. I love my solar panels and marvel at how quiet and efficient they are.

But my wind turbine.... I really, REALLY love my wind turbine. I am a huge disappointment to feminists everywhere because of how much I love my wind turbine. When it's whirring away in a high wind I can watch it for hours. I can go over and stand at the base and gaze up at it. I can put my hands on the tower and feel the energy coursing through the steel and being carried safely by the wires inside into my battery room, where it can accomplish great tasks in the house. We have an AMP meter in our battery room which records how much power the turbine is producing. Our winds are often not consistent, sending the meter up and down constantly, with me cheering when it hits the high end of its capability. In fact, even with a satellite television dish and 100 channels, there is nothing better to watch on a windy night than that AMP meter dizzily dancing up and down.

I share my enthusiasm for my solar panels and wind turbine with you to emphasize a point. For many of us our jobs completely remove us from the possibility of any great sense of accomplishment. Filling out spreadsheets, filing sales reports, pushing bits of information around a computer endlessly can be disillusioning and demoralizing. For what? What is the end result of the investment of my time? Yes, I am earning an income, but if I'm robbing my soul to accomplish it then I'm not going to be very enthused about life.

Choosing to incorporate renewable energy into your home gives you a chance to invest some time in learning about it and then integrating it into your life. That will pay you back in more ways than just a good return on your investment. It will fill your soul with joy as you watch your carbon footprint shrink. It will fill you with confidence that you are making your home and family more independent of outside suppliers of energy who may not be able to provide a consistent product in the future. It will bring joy into your life in a huge variety of ways. But you need to get started, right away.

The first step you should take is to make sure you have some battery power packs around to run your really essential loads. These will include both a battery and an inverter that you can use to run small appliances and charge things like cell phones during power outages. Then you should develop a plan to begin building your renewable energy system. If you have limited funds but want to get started on a system that will be expandable, start with a good inverter. By this I don't mean the kind you can get at a big box store for $200; I mean the kind you'll get through a renewable-energy dealer for $2,000. This more expensive inverter will have features that make it far more valuable than just taking the DC power you have stored in your batteries or that the solar panels make and converting it to high-quality AC. It will have features like "surge capacity." While an inverter may be rated at 2,500W (enough to run your washing machine and microwave), periodically your home will have very large loads. For someone living in the country, for instance, when the pump comes on there will be a huge surge of power required to get the pump moving water. Once the water is flowing it consumes much less power over time. A good 2,500W inverter will have a 6,000W surge capacity to handle those large loads. It may be just for a few seconds, but it will allow your home to function properly when those surges happen. A cheaper inverter is more likely to shut down during a large surge because the load exceeds its capacity.

A good inverter will also be able to charge batteries through your generator. This will ensure that your batteries get the proper charge. If you are connected to the electricity grid a good inverter will be able to handle all the work that must be done on a daily basis. Once the sun comes out it will first make sure your batteries get charged, then once they are charged it will run all your household loads, and if there is more power than your home is using it will divert the excess to the grid for your neighbors to use. On a cloudy day when you turn on the washing machine the inverter will automatically know there is not enough power coming from the solar panels so it will go to the grid to get the electricity you need. A less expensive inverter does not have this capability.

If you invest in a good inverter it will also be expandable. Let's say you spend $2,500 on a good inverter and $1,000 on batteries and $1,000 on solar panels. This will keep the lights on and fridge running during a power outage, and the batteries will be charged when the sun comes out. Next year, when you get your $1,000 income tax refund, you can add more solar panels. The following year you can take that money you set aside in your "Green Energy Bank Account" and buy more batteries to increase how long you can ride out a power outage. Then a few years down the road when your uncle Harry dies, you can take the $5,000 he left you and put a whole bunch more PV panels on your roof. The beauty of investing in a good inverter is that it is expandable and will allow your system to grow without having to be upgraded.

One thing you learn when living with renewable energy is that it takes a huge amount of effort to make "heat," or "thermal" energy. Many people who live off the electricity grid sort of cheat and live more "on propane" than "off the grid." They shift all their largest energy loads, those being thermal loads, to propane, which is liquid natural gas that can be delivered by truck and stored in tanks. Since many off-gridders are far from natural gas hookups, this is convenient. But shifting all your major loads to propane defeats the purpose, especially if you moved off the grid to reduce your carbon footprint or to be more independent. If a lot of your grid-supplied electricity comes from a source like wind power, hydro, or even nuclear, you actually would have a lower carbon footprint using grid power for your thermal loads than propane.

Propane is delivered by truck often from sources far from urban areas, and it could become one of the least reliable energy sources. So trying to find an alternative is important. In our off-grid home we are in the process of trying to eliminate propane. We installed a solar thermal system to heat our domestic hot water. Soon I hope to install a water loop through my woodstove to heat my water as well. This will help on those days when there is no sun. In November and December we often have cloudy days but it's cold enough that the wood stove is going, and it will heat our domestic hot water at the same time.

We have also added more solar panels to produce more electricity. We are doing an increasing amount of cooking with our convection oven, our microwave, and our induction burner stovetop. We also do some of our stovetop cooking on the wood stove and have acquired a wood stove oven that sits on top of the wood stove and can be used for baking. We also have a solar oven, which I strongly recommend everyone have. You can find plans for them on the Internet. After several attempts at a homemade one we purchased one from Sun Ovens (www.sunoven.com). It is a fantastic design and can boil water in 75 minutes when it's below zero outside in March!

We still have the challenge of our propane cookstove. Some people going off grid and off propane use a wood-burning cookstove, which we will move to eventually. With 150 acres of forests, we have all the wood we could ever use. It's renewable, and since it's just releasing carbon it captured during growth, it's carbon neutral. Nothing sounds nicer than making breakfast on a winter morning on a wood-burning stove. Nothing sounds worse than doing it in August in a heat wave, but when that heat wave hits we'll be getting lots of sun so we'll have ample electricity to use all our electric cooking tools like the convection oven and induction stovetop.

As you saw from Bill Kemp's "Renewable Energy Pay Back Time" chart on page 124, when you finally do get around to using solar and wind power for electricity the payback is longer, but there is still a payback. Depending on where you live, what sort of government incentives are available, and what you pay for electricity, the payback time might be anywhere from 12 to 20 years or more. Let's say there were generous incentives in your area and you made an investment of $20,000 with an expected payback of 15 years. Even if it were 20 years, it's still an excellent payback. Most solar panels you purchase today will come with a 25 to 30 year warranty, which indicates that the manufacturer has real confidence that the panel is very robust and will still be functioning long into the future. The original solar panels shot into space on satellites in the 1960s are still working, so there's every indication the panels will be producing electricity for you for decades.

Your investment in solar and wind does have a payback. Now let's look at other investments or purchases you make to compare their payback. What is the payback on your riding lawn mower? What was the payback on your trip south last winter? What was the payback on your $10,000 granite countertop? And what was the payback on your last vehicle? Many people will happily spend $20,000 or $30,000 on a new car or truck knowing full well that in 12 to 15 years it will have no value. There is simply no payback. But for some reason we expect an investment in renewable energy to have a payback. Well it does and it's excellent!

After you hit that hypothetical payback point of 12 or 17 years, you get "Free Power for Life." Over time the efficiency of the panels may decline a bit, but they will keep producing clean, green, emissions-free electricity for your house for decades after they're paid off. Once you get those panels installed you are inflation-proofing your home from rapidly escalating energy costs. Like so much of our society's infrastructure, electrical utilities have not been investing enough to maintain their power grids and generation capacity, and the way they'll deal with this is to pass these costs onto ratepayers. By generating your own electricity you are taking control over something that brings great comfort and convenience to your family. In uncertain times, it will be nice to have lights on, a refrigerator preserving your food, and furnace fans running to keep your home warm.

And what about climate change? Fifty percent of the electricity generated in America comes from coal, which is basically pure carbon that, when burned in a power plant, releases carbon dioxide into the atmosphere, creating climate change. When you use power from the grid you are enlarging your carbon footprint. When you power your home with renewable energy you are reducing it, dramatically. What value do you put on doing the right thing for the planet?

Over the decades that your solar panels will produce energy they will become something that gets passed down to future generations in the family. What price do you put on the creation of a legacy? So many

Americans want to leave a legacy for their children and grandchildren, from family cottages to college educations. How about leaving a family legacy of clean, green, carbon-free energy for future generations? What price can you put on that investment?

I think what has to change is our perception of the true cost of electricity and our responsibility for its impact on the planet. We assume that because it was always cheap and someone else's responsibility we have no obligation to take control of it for ourselves. It's time to change that perception. Stop assuming that someone else will responsibly and reliably provide power to your home. Start creating it yourself. You can start small and work your way up, but you should regard the generation of electricity as your job. You can do it efficiently and cleanly. And it will make your long-dead ancestors proud because you are returning to a time when people had control over their own destiny and hadn't turned over their well-being to an economic entity whose first obligation is to its shareholders, not your family.

It's easy; you just have to make the commitment to take back this control. And it's gratifying. Think of the most gratifying things you've done—your first job, raising your family, coaching a sports team—nothing will be more gratifying than looking up at your roof and seeing those solar panels powering your home, reducing your carbon footprint, increasing your energy inflation protection, and making your family more independent to weather the next storm. Make your grandparents proud! Take back the power. Make it yourself.

14 The Generator

When a hurricane is approaching the TV News people love to head to the nearest big box retailer to film people struggling to fit generators that they've just bought, into the trunks of their vehicles. I'm not sure this is the best strategy for a generator purchase.

Before purchasing a generator figure out what you want to do with it so you have realistic expectations. I watched a news report during Super Storm Sandy that showed a group of people in Manhattan gathered around a tiny generator struggling to get it going. It was outside a restaurant and I got the sense that they were all expecting it to power an entire restaurant, but this generator would have been challenged just to keep a few lights on. I kept thinking there were going to be a lot of disappointed people when they finally got that thing going.

As you've been making your home more efficient hopefully you'll have some idea of how much energy the appliances use. Then you'll need to think about what you want to keep going during a power outage. From this you'll have a better idea of how big a generator you'll need.

Let's start with the best-case scenario, assuming you have lots of money. The best solution to powering your house during a temporary power outage is finding a dealer to install a large, permanent generator. Companies like Generac and Kohler Power Systems make generators designed to power your entire house. They come in noise deadening, waterproof enclosures and have fully automatic controls. They will start automatically and provide power to the house as soon as the electricity grid fails. Once the utility power returns, the unit will reconnect the house to the grid and shut itself down. It's magic! Units can even be programmed to automatically start up periodically, to ensure that they will be ready at the next blackout.

The price on these units will depend on how large your house is and how much you need the unit to power. They won't be cheap, but they provide the most convenient and almost transparent way to deal with power outages. You will need a knowledgeable person to install a unit like this.

Many of us won't have this luxury so we'll have to look at a smaller and potentially portable generator. Generators are usually rated in watts (W) and you'll see them from 1,000 to 10,000 watts. I would suggest you purchase as big a generator as you can afford. If you can lift or a carry a generator, it's too small. The larger the unit the more loads you will be able to run. Most generators won't have any trouble with lights and computers and TVs, but larger inductive loads like electric motors will challenge them. Washing machines and well pumps can be problematic for many small generators and they will cough and sputter and often not be able to run larger loads.

So it's recommended that you test the generator as soon as you get it, rather than waiting for the first power outage. Nothing would be more frustrating in a prolonged blackout than to find the generator you bought can't handle your well pump.

You'll also have to decide how you want to connect this generator. Just running extension cords to your fridge and other appliances is not a realistic solution to powering essential items in your household. Ideally your generator should hook into your home's electrical supply with a transfer switch. This can be manual or automatic, but you should have it installed by an electrician. When it's installed you can select what you think are "essential loads." These would be things you want to keep powered during the blackout, such as lights, the fridge, the freezer and some wall outlets to charge cell phones and keep a TV and computer on so you can monitor what's happening. In the country your well pump would definitely be essential. If you have an electric stove this would be non-essential because most small generators couldn't handle it. You can go a few days without the washing machine and dishwasher so they would be non-essential as well.

If you heat with natural gas, unless there is an earthquake, it is likely that natural gas will keep flowing your home, but without electricity you will not be able to power your furnace and furnace fan to force the heat through your home. So identifying your furnace as an essential load would be critical as well. A furnace will be a fairly large load so it's another reason to be looking at a 7,500-watt or larger generator.

When the electrician installs your transfer switch so the generator can power your 'essential' loads, you should have him provide the wiring required to the generator. It will be hard wired into the transfer switch and then go to where you plan on running the generator. If you live off-grid and run a generator on a regular basis you may want to consider building an actual generator building. Or if you invest in a larger, heavier generator you'll have to find a permanent location to install it. These would be hardwired, or permanently connected from the generator to the transfer switch.

For a smaller home emergency system you probably won't want to go to this expense. So decide where you'll be running the generator. It has to be outside, so pick a place as close as possible to your electrical panel box. Then the electrician can make up the cable to run to the generator. Most generators will come with a heavy-duty "twist lock" connector to plug into.

The wire or cord that will run from your panel box to the generator is very important. It must be sized correctly to ensure you get the maximum output from your generator. The first time I had an electrician familiar with generators come to check out our off-grid electrical system he pointed out the "extension cord" that ran from our generator to our inverter to charge our batteries was drastically undersized. Once he installed a proper permanent wire we went from getting 30 Amps of charging power to 120 Amps going into our batteries. That undersized cable was only letting 25% of the potential energy from the generator through.

So this is why I recommend you get an electrician involved in the installation of your generator. While the salesperson helping you at the big box store may try to convince you that the heavy-duty power cord supplied with the generator is sufficient, they are probably not an expert in this area.

The transfer switch and the separation of essential and non-essential loads needs to be done in such a way that it meets the electrical code. Having your neighbor 'who does this stuff all the time' help you do the install is just not a good idea. You have to remember that just because you're dealing with a generator or battery bank and inverter, you can still be killed or seriously injured, just as you could with the energy from the grid. You also have to remember that improper wiring can lead to fires and other hazards. The worst plan for an extended power outage would be to have a fire at your house. Fire and emergency crews are going to be really busy, so the likelihood of them responding to your call is going to be severely reduced. So don't turn a bad situation, like a power outage, into a personal disaster. Get it done right. Find an electrician who can recommend the proper generator and wire it properly up to code.

Fuel

You'll also have to give some thought to how you power that generator.

- **Gasoline:** Everyone is familiar with the small gasoline generators that are common throughout North America. Cheap and easily fueled, they have a very short life span when used in demanding applications. The majority of gasoline engines operate at 3,600 rpm, which results in rapid wear and high noise levels. For off-grid applications, expect a life span of five years or less before a major rebuild is required. But as a source for power during an emergency they can be (on a small scale) sufficient.

- **Natural Gas:** Natural gas engines are offered in two varieties: converted gasoline and full-size industrial. The converted engine is really no better than a gasoline engine, except that it offers the advantage of no fuel handling because it can be connected directly to the gas supply line. Industrial-sized natural gas engines are of a heavier design and operate more slowly, typically at 1800 rpm. This increases engine life and greatly reduces engine noise. However, natural gas is not available in all areas because the fuel is transported via pipelines.

This becomes a tough decision for someone with a less than positive view of the future. How long will natural gas keep flowing to heat my house and power my generator?

- **Propane**: Propane is similar to natural gas with the exception that this is the fuel of choice for off-grid and rural applications. Propane may already be the fuel source for other appliances in your home, making generator connection a breeze and eliminating an additional fuel source at the home.
- **Diesel:** The diesel engine has the best track record for longevity. Diesel units are heavy, long-lasting machines that generally operate at low speeds. Fuel economy is highest with a diesel engine. The downside of diesel generators is that they tend to cost more and have a high operating noise level, necessitating a sound-damping enclosure. Diesel generators are the generator of choice for many off-gridders.

If you go with a gas or diesel generator, you should have a number of jerry cans of fuel on hand. The challenge is that neither of these fuels will last forever. Over time the fuel will become unstable and will not work in the generator or your car's engine. So if the fuel is going to sit for a while, add fuel stabilizer to it. This should allow you to store it for one to two years. I would suggest that you not store it that long, and use it in your car or lawn mower from time to time and replace it with fresh fuel. Every three to six months use the stored gas in your car and refill the storage containers. Store the gas in a cool location and out of direct sunlight.

If you have a diesel generator make sure you purchase the yellow plastic containers to differentiate it from gas. Put a tag on the jerry can that indicates when you purchased the fuel and whether or not you've added fuel stabilizer to it. Fill jerry cans as full as you possibly can. Condensation in the unfilled area will be a problem for stored fuel so the less air in the tank to condense, the better. And remember, you have to be careful where you store a flammable product like this. Do not store it anywhere near a source of heat or ignition. Don't store it inside your house. Never fill your generator in a garage or enclosed area because of fumes or the risk of a spill. Don't smoke near these containers.

The key to a generator working well is to run it periodically. Too many people have generators that sit in a garage for years, and then when the power goes off they can't get them started. With any internal combustion engine you need to use it periodically to keep it in working condition. If it is an electric start you'll have to make sure the battery stays charged as well. You should put a small "trickle charge" battery charger on it to keep it charged. You should also have jumper cables so that you can jump start the generator from another battery if you can't get it started when you need it. Electric starts are an excellent option on a generator because pull starts tend to be fairly difficult, or at least require a pretty good yank. So if you have back problems or a spouse who doesn't want to have to struggle with the pull start while you're away at work, get an electric start, and keep that battery charged and ready to go. If it's a smaller electric start don't assume that the generator will charge the battery when it's running. Keep your eye on the battery voltage to make sure it's high enough. Start the generator every two or three months to make sure it's ready when you need it.

Having a generator for temporary power outages is a great idea. Make sure it's wired and connected properly and that you test it regularly. Also make sure you have lots of fuel around to power it. You'll be surprised at how quickly you'll burn through fuel in an extended blackout. Which brings us back to the question, well, what if the grid takes a really long time to come back on, as more and more customers are discovering after these more intense storms that really tax utilities? This is where a renewable energy strategy for your home will help. Having some batteries and solar and wind to charge them is part of a strategy to keep your home powered and life as close to normal as possible for as long as possible. Ideally you would be able to run your home off your batteries without the generator, and rely on the generator only if you have several days without sun and wind. This is the strategy of people who live off the electricity grid and one a sensible prepper should move towards.

If ultimately this is your goal then I highly recommend *"The Renewable Energy Handbook"* by William Kemp. It expands on all the information I've discussed over the last few chapters and is the best resource available for a detailed and understandable approach to energy independence.

15 Food

According to the U.S. Dept of Health and Human Services a moderately active adult needs about 2,000 calories a day to live healthily. That's pretty easy these days. In fact, it's so easy that a lot of us consume substantially more than 2,000 a day. So how and what you eat is going to have a great impact on how you weather the changing world we live in. The one defining theme should be for you to "eat lower on the food chain."

The planet now has 7 billion people on it competing for food. Most of us are aware that it takes more resources to produce meat for humans as opposed to just consuming the grains directly. So one strategy for a future on a planet with more competition for food is to adopt a more plant-based diet. It is also much easier and cheaper to store items like rice and pasta and beans, than it is to store meat.

In the spring of 2008 as oil rocketed to $147 a barrel, the world experienced its first food shock in a long time. The price of all commodities went up and this included food products. Rice in particular became very expensive and in short supply. Asian countries that used to export rice began saving it for their own populations, which compounded the problem. As North Americans of Asian descent started realizing that there were shortages of rice elsewhere in the world they started purchasing extra, which caused supplies on the west coast to grow tight. Some stores even implemented restrictions. Big box stores in particular looked at the purchasing history on loyalty cards and if you didn't have a history of purchasing three 20-pound bags of rice a week, they weren't going to let you start to do that during a shortage. This is one of the downsides of using stores that track your purchases.

It's a good idea to start building a supply of food in your home to help you the next time there are shortages. Many experts in the food industry are suggesting that it isn't **if** it happens again, it's **when**. A root cellar is an excellent place to store vegetables, but it wouldn't be recommended for most of your food. You should start with a "pantry" or place to store dry and canned goods. It should be cool, dry, and dark. If you have a large closet that's not being used it's a good place to consider for your pantry. If all your closets are full it's time to pick the one that's the most convenient and best suited for food storage and have a garage sale with whatever is in there now. Storing food in the basement is only a good idea if it's not too humid down there. Basements tend to be cooler than upper floors but the humidity will definitely affect how well some foods last. While canned goods in a damp basement will store fine, you may find that the labels get moldy and have to be removed, which is inconvenient to say the least. While you may have an extra closet on the top floor of your house, it may be too hot to use for food storage. Things like pasta won't be affected by the heat too much, but over time the foods stored in cans and any food with oil in it, like a rice kernel, will deteriorate faster. So the ground floor is best if possible. The closer to the kitchen the better.

The average temperature in your pantry should be above 32°F (0°C) and below 70°F (21°C) but remember that the cooler the storage area, the longer the retention of quality and nutrients. It should be dry, with less than 15% humidity, and food should not be stored on the floor. Try and have the lowest shelf two feet off the floor. You should also try to make sure insects don't get in, and have a mouse and/or rat trap if you don't think you can keep rodents out. If you have an abundance of large plastic storage containers, putting the items inside these is an excellent defense against pests, especially with things that are easily chewed into like pasta and rice that come in bags. It's also best not to have electrical equipment such as freezers, furnaces, and hot water heaters where you're storing your dry goods because they will produce heat, which you want to avoid.

Start with canned goods and things you know will keep. Pasta in bags will last a long time. Rice as well. Canned goods should last years. As much as possible you should avoid packaged goods like cookies and crackers because they'll get stale fast. You should start by picking up things on sale. If you're at the grocery store and pasta and canned pasta sauce are on sale, buy what you would normally buy for your day-to-day consumption and then buy two or three more for storage. Or ten more! Look for specials on canned soups, canned fruit, canned vegetables, dried beans and lentils, and bagged rice and pasta. These can form the basis for your pantry. Then start to add those things you think will help enhance your menus, like sauces and spices, which last fairly well. Include non-essentials that you enjoy like salt, sugar, tea, and coffee, because they'll be nice to have if there are shortages.

Shopping like this will seem strange at first. Many of the things you'll be stocking up on are the sorts of things you might have purchased when you were filling up an extra bag during the holidays for the local food bank. You'll be buying things you might not regularly eat, such as canned fruit and vegetables. We've had the luxury of fresh produce 12 months of the year for so long that it's hard to remember ever having had to rely on canned versions of these. I'm not suggesting that you're necessarily going to shift your diet to canned versus fresh vegetables; I'm recommending that you start building up a three- to six-month supply of food that you have access to if you need it. It doesn't necessarily have to be a food-shortage crisis either. It can be part of your plan in case you suffer the loss of an income. Part of your defense plan can be to dip into your pantry during the transitional time until you can replace that lost income. Remember, the strategy is to start filling up the pantry with items on sale. So not only is that meal of rotini and a wonderful jar of tomato basil sauce less expensive than the fresh version and way cheaper than the restaurant version, since it was also purchased on sale it is even more affordable. You're protecting yourself not just from food shortages but from income disruptions, so stocking up the pantry is an excellent financial strategy.

Once you get your pantry stocked up you can start a maintenance program. As you add foods, label them with the date of purchase. Then, next time that item is on sale again, you can take the older item out to consume soon and put the newer item in. This is the same strategy that grocery stores use. They pull the older inventory to the front of the shelf and put the newer stuff in behind to make sure that the older items sell first. Accountants call this the "First In, First Out" (FIFO) system of inventory management.

How long the items are going to last in your pantry is going to depend on how cool and dark and dry your pantry is. It's also going to depend on your comfort zone. We're all used to the "Best Before" dates and "Expiry Dates" on some items we purchase. Some items like canned goods will have a "Packed Date" which will tell you when it was actually put in the can. Most of us have eaten items well past their "Best Before" date. They may not have been as great as when they were fresh, but they are fine. I would tend to be more careful with meat and dairy products.

The following list deals with emergency storage for situations like a hurricane or severe snow or ice storm. I have not included readily perishable items that you would store in your fridge like meat and dairy products. This is for your long-term pantry storage.

Foods Recommended for Storage

- Water: one gallon per person per day for drinking, cooking, and personal hygiene
- Ready-to-eat canned foods: vegetables, fruit, beans, meat, fish, poultry, meat mixtures, and pasta
- Soups: canned or "dried soups in a cup"
- Smoked or dried meats, commercial beef jerky
- Dried fruits and vegetables: raisins, fruit leather
- Juices: vegetable and fruit, bottled, canned, or powdered
- Milk: powdered, canned, and evaporated
- Staples: sugar, salt, pepper, instant potatoes and rice, coffee, tea, cocoa mix
- Ready-to-eat cereals, instant hot cereals, crackers, hard taco shells
- High-energy foods: peanut butter, jelly, nuts, and trail mix, granola bars
- Cookies, hard candy, chocolate bars, soft drinks, other snacks

The following chart from Colorado State University gives approximate times for food storage: (www.ext.colostate.edu/Pubs/emergency/fdsf.html)

Optimum Length of Storage for Quality and Nutrition

Fish, canned	18 months
Canned potatoes	30 months
Dehydrated potatoes	30 months
Canned fruits and vegetables	24 months
Canned fruit juice	24 months
Canned vegetable juice	12 months
Pickles	12 months
James and jellies	18 months
Rice, dried	24 months
Cornmeal	12 months
Pasta, dried	24 months

Cold breakfast cereal	12 months
Prepared flour mixes	8 months
Packaged dry beans, peas, and lentils	12 months
Canned evaporated milk	12 months
Dry milk products	24 months

Canned foods keep almost indefinitely as long as cans are undamaged. The question is the quality of the food when you finally open it. In certain situations most of us would be quite happy to have some food that is safe, even if it's a year or two past its optimum storage time. I would suggest that the optimum times listed above are a best-case scenario for maximizing the nutritional value of the stored food. If you needed to eat and the food in the can had been there for four or five years, you'd probably be happy to eat it. This is a "beggars can't be choosers" scenario. The fact that most guides suggest that rice and dried pasta last two years tells me it will last much longer than that. It will simply be important for you to have a system of labeling so that as you put new items into the pantry you can take older ones out. I recently ate some canned mandarin oranges I put into my pantry four years ago and they tasted great. I believe that much of the labeling you see in terms of the storage life of products is based on product liability, where the producer does not want to risk someone getting sick from a product that has been stored too long. The longer that product sits unused, the greater the likelihood that it will experience less than optimum storage conditions like extreme and extended heat waves that will accelerate its deterioration. I also think that food manufacturers would rather you turn food over more regularly so you'll have to purchase more of their product.

From a cost-savings standpoint you should also look at purchasing some of the items for the pantry in bulk. This will save you money but will require a bit more time and possibly some expense in terms of how you store it. The advantage of purchasing things like flour and rice at a grocery store is that they come in a fairly well made bag. If you buy in bulk you'll have to decide how you're going to store it. Mason jars are an excellent way to store things like dried fruit and beans, but you can avoid this expense by just washing glass jars that you would normally recycle. Start looking at what you're recycling and see if you're actually throwing away excellent storage containers. Plastic peanut butter jars are good because they have a screw-on lid that creates a fairly airtight seal and ensures that pests stay out. My favorite containers come from the brand of Caesar salad dressing that we purchase because they're glass but have a great plastic lid. With scary chemicals like BPA (bisphenol A), which is an endocrine disruptor found in plastic, using glass wherever you can is a good idea, especially if what you're storing in it is acidic and may tend to leech chemicals out of the plastic.

So get creative. If you're buying in bulk to save money you'll need to think "outside of the box" (since it won't come in a box) to come up with the most cost-effective way to store things. If you live in a neighborhood with curbside pickup of recyclables, once you find that favorite plastic peanut butter jar or whatever works for you it may be worth a quick walk around the block after dark or early in the morning to load up. Free storage containers! How great is that!

Fresh fruits and vegetables are extremely healthy and a huge luxury when you're eating them out of season. One of the keys to saving money and preparing for a time when there will be less diesel fuel for trucks to haul fresh produce from the south year round is to start to eat in season. Most of us have come to realize that a strawberry picked early in California or Florida and shipped to a place where there's snow in February is not as tasty as one picked at the height of the summer and shipped a short distance fresh from the local farmer. While the quality of those winter strawberries is getting better and better, is it really normal to eat strawberries in the winter in the north?

I would suggest one of the best ways to prepare for the inevitable changes to come would be to start eating as our ancestors did: in season. That means having strawberries every day for four weeks in June until you're sick of them. Then you move onto raspberries, then peaches, then apples in the fall and so on. By the late fall you'll be pretty much having just apples and pears, and as winter sets in your apples will be in pies and sauces.

It's probably time to start working on your canning skills. Many of us grew up with parents and grandparents who canned, but as we got busier with two-income families, and as lower energy costs made buying food so cheap, it didn't make much sense to can anymore. But those days are over. Canning and preserving are lost arts which you'll have to work on to get good at. Canning also adds a certain degree of danger to your food because if you don't do it properly your canned fruits or vegetables can cause botulism. While many of us are getting quite comfortable with the idea of injecting "botox" into our faces, eating food contaminated with it is another game altogether and one that can be fatal.

There are some excellent books about canning and preserving you may want to start with. You may also want to find a local course on the subject. It's always easier to learn from someone who's good at it. Maybe it's time to have grandma come and stay at your place for a couple of weeks at the peak of canning season and you'll have an in-house coach at your disposal. There's nothing like experience in this department.

If you are going to start canning consider using a pressure canning system. Our experience with canning has been that it is extremely energy intensive. The amount of energy you use for boiling, whether it's electricity or natural gas, is quite staggering. I marvel when I'm at the grocery store at the price of food, especially canned goods. Once you've done it yourself you get an appreciation for how cheap energy has contributed to our high standard of living. The company doing the canning on your behalf has to purchase energy for the process as well as purchasing the fruits and vegetables to begin with, along with cans or glass, labeling, and cartons for shipping. The list seems endless and the final price to the end consumer mind-bogglingly cheap. This is just another way in which my research into peak oil and energy has made me grateful that I live at the time I do. Walking up and down a grocery aisle has become a time of wonder for me. Michelle eventually just ignores my ranting: "How can they possibly make a can of beans so cheap!" Now that I think of it, it's not just in the grocery aisles that she's stopped listening to me.

Using a commercial pressure canner ensures that you get the temperature to 240°F (116°C). This is the minimum temperature required to destroy botulism spores. Like a pressure cooker, a pressure canner will reduce how much energy you have to put into the canning process, which will save time and money.

Many of the books that have been written over the years for back-to-landers, those who want to get back to the simple life, always stress canning as a way to save money. Lots of money. I'm not really convinced in this day and age that that's the case. I won't argue that it might come to pass, but when I look at how long it's taken us to can things and how much propane we had to burn to do it, I'm simply not convinced that there are huge cost savings today. This is why I emphasize stocking your pantry with commercially canned goods now. You know they have been canned properly and someone else has had to spend all the money on the energy to do it. I think it's good to have the pressure canner and experiment with it to develop the skills to do it in the future when you have to, but as to it offering significant cost savings right now, I'm not convinced we're there.

A freezer is an excellent alternative or supplement to your canning. Many years ago we invested in a 10-cubic-foot freezer. While this may not sound like a big step, for someone living off the electricity grid it is. We had to evaluate how much additional load the 353 kilowatt-hours per year represented and determine if we could afford the electricity from the finite amount we produce with our solar panels and wind turbine. We had recently upgraded to more solar panels so we felt we could. We also decided to put the freezer in our basement. Our house, which was built in 1888, has a concrete basement, which is not heated. So it's more like a crawl space and stays very cool in the winter. Our feeling was that since the air temperature is close to zero the freezer wouldn't have to work that hard. And so far it appears to be the case. We haven't been able to notice any significant difference in our electricity loads.

If you were on the grid and paying 12¢/kWh, the electricity to run the freezer would be $42 (353 kWh x .12¢ = $42.36). We really loaded our freezer up with vegetables last summer. We froze beans, peas, basil, cauliflower, broccoli, and tons of tomatoes. Now on a cold winter day we take out a bag of tomatoes, put them on the woodstove in the morning, throw in some garlic and basil, and let it simmer all day, filling the house with wonderful smells. And their amazing flavor is like having a bowl full of summer. The tomatoes didn't go through a lengthy heating process by being canned; they were just put in bags and frozen, so I think we preserved some of the healthful benefits of freshly picked tomatoes.

Another process you might want to look at for preserving food is dehydration. Dehydrating food can be as simple as slicing it and putting it in the sun on a warm day. What you are attempting to do is remove the water through evaporation. The warmer and drier the air the better. To speed the process you may want to look at building a solar dryer. We have a solar oven, and during the hot sunny days of August we often don't need it since we have more than enough electricity for all our cooking. So we started putting tomatoes in it and it was amazing how quickly they would go from the garden to "sun dried," usually in one hot day. I also used one of my original solar oven boxes to supplement my output. The challenge with these is air movement, since you want to remove that moist air and keep hot dry air entering. Most models have a central drying shelf with a glass cover that creates a greenhouse effect, with a vent system at the top and bottom to bring hot dry air in at the base and vent the moist air out the top.

As energy and food get more expensive it's going to make more and more sense to take the time to build a solar food dryer and dehydrate fruits and vegetables from your garden. You will need to determine when the time is right for you. Building a solar food dryer may only require $50 in materials, but it may take a couple of days, or even a week if, like me, you're not a natural-born carpenter. If you still have a full-time job where you earn a wage, that investment in time may not make sense. If you are working part-time or are between jobs, it might make complete sense to build your own food dehydrator. Most people with gardens know that when the harvest comes in you'll inevitably find that you've planted more food than you can consume, so it just makes sense to have a low-energy way to save it.

You need to come up with what works for you in terms of how you save and store food. Find the best combination of purchasing commercially produced products, canning, freezing, or dehydrating produce from your garden, or buying from a farmer in season when food is at its cheapest. As food and energy prices change and your employment and income situation changes, you'll have to find the right combination that works for you at any given time. I would suggest you have the equipment and skills you need to exercise any of these options to varying degrees as things change and evolve in your life.

The "no-outside-energy-required-at-all" storage option for some fruits and vegetables is a root cellar. This may sound like a very old and rural concept, but a root cellar is an excellent, low-energy, very cost-effective way to store produce. Whether it's the bounty from your backyard vegetable garden or food purchased from local farmers in season, a root cellar is an excellent way to save money, become more independent, and reduce your carbon footprint. Potatoes and apples grown locally and stored with no refrigeration have a much smaller impact on climate change than those driven across the country or flown in from New Zealand or South Africa.

A root cellar can take many forms but the key is that it be cold and dark and humid. Ideally the temperature should be just above freezing to about 40°F (4°C). The humidity should be from about 80 to 90%. Knowing how dry the air in your home gets in the winter, you can see this is going to be a bit of a challenge. There are some excellent books listed in the appendix on building a root cellar. One option is to build it outside by digging it into a hill or by simply digging a hole, lining it with concrete blocks or stone, putting a roof on it, and covering it with soil. If you're living in suburbia this may not seem like a normal thing to do, but sometimes desperate times call for desperate measures. I can't think of a better way to get to know your neighbors than when they come over to ask you if you're putting in a pool or if that's a 1950s-style bomb shelter you're digging.

Older homes often have unheated basements which are an excellent starting place for a great root cellar. A dirt floor is even better, not only because the floor will get lots of dirt dropped on it anyway but because it will help regulate moisture. If it appears that the cellar is getting too dry in the winter, you can simply pour some water in the dirt so that it can evaporate out later. You can also do this on a concrete floor or use a bucket of dirt or gravel to get humidity into the air. If you have a finished basement then you need to build a small room, preferably on the north or east wall where it will be coldest and preferably with a small window. Or if you have an unfinished basement that you may finish, when you put in the root cellar insulate the walls to the inside to keep the heat out and leave the outside walls uninsulated to allow cold air to enter. There should be no ductwork or heating sources. The goal is to provide a way for warm air to leave the root cellar and cool air to enter. If your basement is naturally cool, leave one opening low on the basement side and place one vent high in the outside wall for warm air to leave. This way the cooler air will have to travel across the root cellar and create some air circulation.

The root cellar needs to be dark to prevent potatoes and onions from sprouting prematurely. Try and seal up all the cracks to keep out rodents and pests, and make sure you have a couple of mouse traps baited at all times. Don't store any of your canned goods or preserves in the root cellar because it will be so moist it will cause any metal to rust. One compact fluorescent light bulb is all you'll need for light when you enter the room, and make sure you use a ceramic base and outside wiring because it will be cool and damp. If the room has a window you'll want to cover it with a heavy blind to keep the daylight out. One nice design I've seen takes the space of the window and has it half covered with a plywood vent that extends all the way to the floor and brings the cold air down to the floor of the root cellar. Above that is another vent for the warmer air to exit, keeping a good circulation going.

Next you'll want to build shelves to hold all those fruits and vegetables. Some people like to put things like potatoes and apples in wooden crates. You can store potatoes in dry sand or peat moss in plastic buckets. I've found that wire mesh waste baskets are excellent for onions because the air circulates completely around them and I can hang the baskets up to get them off the ground. Squash and larger vegetables can just sit on shelves. While root crops like potatoes like to be stored in a dry medium, carrots like a bit of moisture, so I store them in damp sand or peat moss. One of the keys to maximizing how long items will store is to ensure that you pick the best fruits and vegetables at harvest time and handle them as carefully as you can. Bruises and blemishes will hasten rotting and you'll often notice that this is where mould starts.

When you're harvesting potatoes or sorting apples, store the nicest-looking ones with the least signs of damage, and eat the other ones first. As the fall and winter progresses keep your eye on things and try and eat the items that look as if they may be starting to go. This is always is a good excuse for a nice soup or stew. If some of the skins on the potatoes aren't looking great just make a big pot of mashed potatoes rather than baking them.

Different fruits and vegetables are going to have different optimum temperature and humidity levels, and you cannot always control these perfectly. Every year will be a different experience with your root cellar and you just need to go with the flow. Ideally you should be harvesting potatoes when it's dry outside, but if August turns out to be wet you may not have a choice. That year your potatoes and turnips may not last as well, but the squash and apples will do great. Your diet will start to echo the health of what you've got stored. Late in the fall you should eat more of the items that won't last as well like leeks, turnips, and beets; well into the winter you'll still have lots of potatoes, carrots, onions, and some apples.

The majority of the vegetables I store in my root cellar are potatoes. I love potatoes and they are easy to grow and store and provide excellent nutrition. If you had to you could pretty much live on potatoes. Oh sure it would get a little boring, but the reality is that you can grow a good chunk of the food you need in a suburban backyard if you concentrate on potatoes.

Potatoes are rich in carbohydrates and protein, which makes them an excellent source of energy. The amino-acid pattern of the protein in a potato is well matched to what humans need. Potatoes are very rich in many minerals and vitamins, providing one-fifth the potassium requirement, and are particularly high in vitamin C. A single medium-sized potato contains about half the recommended daily intake of vitamin C, so if you're having a crisis of conscience about that morning glass of orange juice that is trucked from the south to your breakfast table, don't worry; the potato has you covered. The United Nations declared 2008 the

International Year of the Potato because, *"The potato produces more nutritious food more quickly, on less land, and in harsher climates than any other major crop—up to 85 percent of the plant is edible human food, compared to around 50 percent in cereals."* Take this to heart and start making the potato a big part of your future food strategy.

Your root cellar will help to convince you just how amazing nature is. Our root cellar is actually the old water cistern under the kitchen. It has one-foot-thick concrete walls and stays just above freezing all winter. It is completely dark and yet by late March the potatoes are starting to sprout eyes, the onions are starting to send up green shoots, and the carrots are sending out feathery green growth on top. Somehow, even in the cool inky blackness, they know it's time to start the cycle all over again. And when those potatoes get to the stage where they're too wrinkly and the eyes are long enough, I'll take them out to plant in the garden and each potato will produce another eight to ten potatoes for next year. Free potatoes! After the initial purchase you'll have free potatoes for life. We continue to add new breeds just to keep a good variety to help deal with varying conditions, but growing much of your own food and storing it and harvesting the seeds for next year allows you to significantly reduce your food bill and makes you much less dependent on others.

I cannot emphasize enough the importance of having your own garden, but depending on space and other factors you may not be able to grow as much as you'd like. An excellent option is to join a CSA or Community Supported Agriculture. In a CSA you and a number of other families purchase a share in a farmer's harvest, in advance. I know, it sounds crazy, paying for something you don't have yet. But it takes away so many of the variables that make it tough for small farmers to make a living. It gives them money up front for seeds and they get paid for their time rather than having to wait until they sell the harvest. You as a member of the CSA also have a greater stake in the actual process of growing food. Your share of the harvest depends on what grows well. Some years you'll have tons of some items and very little of others. This is how it is when you grow food. Conditions change. Some vegetables do well, some don't. The megasuperstore has given us the mistaken belief that every shelf is always full of every possible fruit and vegetable. Cheap oil has allowed us to ship produce from wherever in the world it is growing well, but that's about to end.

So belonging to a CSA lets you experience a diet like the one your grandparents had. It also gives you a connection to the most important person on the planet. Not a lawyer or accountant but someone who actually grows food, something we're all pretty dependent on. CSAs have created a food production model that is also helping many younger farmers get into the business of growing healthy, local food. The traditional agricultural model requires massive capital to purchase land, equipment, and fossil-fuel inputs. The CSA model allows someone with a passion to grow food sustainably to earn a living. Some CSA farmers will also require or provide the option of working on the farm to offset some of the expense.

If you can find a CSA like this, join it. Set aside every other weekend or some weeknights or however much time you can afford to work with this farmer. First, there is nothing like the joy of hard labor nurturing food. In this case you'll get the added bonus of learning from an expert. It may seem as if you're just being assigned tasks and following orders, but you'll be learning what goes into growing food and there's no better way to put the joy back in a meal than to understand the toil and sweat that went into it. And make sure you make a nuisance of yourself picking the farmer's brain as to what she's doing and why. Why am I planting the beans here? Why am I watering the carrots and lettuce but not the garlic? The more knowledge you can soak up from this person the better your own garden will be. You'll also be adding to your knowledge base in case this is something you want to try yourself. There is one new reality in the economically and fossil-fuel challenged future and that is that more of our time and resources are going to be devoted to growing food. More human capital, more human labor, and more human time and ingenuity, because we simply won't have the gas and diesel to displace all the labor. Becoming an expert at food production will be a great skill set to have in the future.

The website *Local Harvest* (www.localharvest.org) is a great resource for finding a local CSA or farmers' market in your area. The site also lists local farms and Co-ops in your area. Food co-ops are an excellent way to save money on your grocery bill. You belong to a buying group, which increases the group's purchasing power and reduces its costs. Co-ops traditionally have been a means for people to purchase organic food, which in the past has been much more expensive than traditional food. Often you'll buy cases or larger quantities, which you may want to get other members of the group to split with you. Food co-ops are an excellent way to try and cut out the middleman of the grocery chains and get more money to farmers and smaller producers. They can also save you money on organic foods. According to the USDA, in the current economic model farmers end up with less that 20 cents out of every dollar spent on food; any way you can increase that is good for sustainable agriculture.

As discussed in the "Where to Live" chapter, one of the advantages of living in a densely populated city is the opportunity for food to be delivered centrally. So make sure that if you live in a city you're well versed in where and when all the farmers' markets are. Start to frequent them and make sure you take the time to get to know the farmers you're purchasing food from. Find out where they farm and what kind of farming they do. You may also want to find out if they'll sell to you at better prices if you order more or if you pick it up. It may be worth a drive out to the countryside in the fall to load up your car with potatoes and apples and carrots and squash and all those things you need for your new root cellar. If the farmer doesn't have to haul it into the city she may be happy to let it go at a better price.

If you continue to have animal products in your diet, keep in mind that smaller farmers are much more likely to treat their animals well. The agri-food industry has forced many farmers to grow food on such a scale as to require animals be kept in confined areas and often have antibiotics routinely added to feed to reduce the potential for sickness and economic loss. Smaller farmers may not have the capital for that or may choose to operate at a smaller, more humane level. You may pay a bit more for their meat and eggs and dairy products, but at least you know that the suffering of an animal isn't part of your next meal.

The other huge benefit of getting to know farmers is that in the case of disruptions to the food supply, be they man-made or natural, it's always nice to have a connection to someone who is in the business of producing that which is essential for life. It's unlikely that the chain store will really care if you can't get enough rice or other essentials for your family. That's not the capitalist model. The network of local farmers you've come to know will probably be much more compassionate in the same circumstances. They may suggest that they need your labor more than they need your paper dollars in trade, but so be it. It may be a nice distraction from everything else that's going on to get back to the simple and joyful task of growing food.

One final option that you may want to consider for your pantry is freeze-dried foods, which will last much longer than most of the other techniques I've discussed. Freeze-dried foods are sealed to prevent the absorption of moisture and can be stored at room temperature for years.

With the very low water content the microorganisms and enzymes that would normally cause food to spoil and degrade can't do their work. When you add water and cook them the flavors and smells are excellent. Anyone who has camped and taken freeze-dried food knows how wonderful a nice pasta primavera meal is after a day of backpacking or canoeing. It's like having a restaurant-quality meal in the middle of nowhere with a minimal amount of preparation.

The downside to freeze-dried food is the cost. The freezing, primary drying, secondary drying, and packaging require lots of energy and lots of time, so you have a corresponding increase in cost. Many people who sense a "hard landing" ahead see freeze-dried food as the ideal long-term food storage strategy. The food will last a very long time and because of the low water content it tends to be fairly light and portable. Mountain House claims their pouches of freeze-dried food last 10 years. They also store some of their freeze-dried products in 10 lb. cans and they claim a 25-year shelf life. They even claim they've tasted food from cans that were 35 years old and the contents still tasted fine.

So if you can afford it or just want to invest in a good supply of food and put it away and not have to think about it again, freeze dried is the way to go. Companies like Nitro-Pak (www.nitro-pak.com), Mountain House (www.mountainhouse.com), and Harvest Foodworks (www.harvestfoodworks.com) are a few of the companies that sell freeze-dried food. Be forewarned that there can be a fairly strong survivalist theme in discussions of freeze-dried foods so you may want to refrain from bragging about your new acquisition at your next party. Although I will wager that if you bring it up you'll be swamped with questions from those who've been toying with the idea! Just remember to keep these foods in a cool, dark and dry place to maximize their shelf life.

For anyone with a newborn or expecting a newborn I have not provided any guidance about what formula to buy and how to store it because I believe the old adage, "breast is best." There are a myriad of benefits to breastfeeding your baby on top of the fact that it's cheaper and storage isn't an issue. As long as mom is eating a healthy diet, baby is looked after. From an environmental perspective it reduces packaging and the energy costs of producing and shipping formula. From a health perspective it has been shown to be infinitely better for babies. As is so often the case, returning to the way our parents or their parents did things is a better choice than those provided by corporations today.

(As an aside, which doesn't have anything to do with food but is another example of going back to the "old" ways of doing things, we purchased 24 cloth diapers for our daughters when our oldest was born. Twenty months later when the second came along the eldest was just about toilet trained, so both the girls used the one set and they're still useful as rags in the garage. Think about how many trees weren't cut down and pulped and shipped and processed and used once to end up in a landfill. Disposable diapers have a hugely negative effect on the planet and on your pocketbook. Ask for cloth diapers at the baby shower; then calculate how much you would be spending on disposables each week and take that money and set it aside. I'll bet you can purchase a significant number of photovoltaic panels with that money. Independence often involves choosing what's right for the planet and your pocketbook, not what's most convenient.)

So I recommend you get a pantry stocked with six months' or a year's worth of food and when you can, do it with items that are on sale. This makes for a fallback in the event of job loss or food disruption. Then build yourself a root cellar and stock it fully every fall. If you can grow the fruits and vegetables you store, excellent. If you can't, join a CSA or get to know some local farmers who'll sell you large supplies in the fall to help you stock the root cellar. Yes eat fresh food as much as you can, but having a supply on food on hand just makes complete sense in these times.

16 Gardening

The typical North American diet is one rich in fossil fuels. From the natural-gas-based fertilizers to the diesel used in tractors for planting and harvesting and in trucks for transport, to refrigeration and packaging, the fossil-fuel inputs into our diet are enormous. Many geologists warn that the world has hit peak oil, biofuels are competing for grains such as corn and soybeans, and climate change is causing droughts that are hampering grain harvests, driving stocks to their lowest levels in years. As all of these variables drive up the cost of producing our food, you should plan on spending more of your paycheck at the grocery store.

There is one way to offset some of this food sticker shock, and that's to start growing some of your own. Rule number one for a prepper is to learn to grow food. Whether it's in your backyard or at a local garden rental plot in your city or in some containers on your balcony, it's time you took control of what's on your dinner plate. Gardening is phenomenally popular, but most North Americans focus on flowers, and while some flowers are edible, they don't add to your family's food security or help your monthly budget.

Tremendous media attention is given to the concept of eating more locally, like the 100-mile diet, where you try and offset some of the miles your food travels. The estimates vary, but you'll find that your average North American meal has traveled between 1,000 and 2,000 miles. Some estimates have it as high as 2,500 miles, which has a tremendous impact on the planet in terms of the greenhouse gas emissions created to get it to your plate. For much of the year fresh food being eaten in the northern States and Canada has had a long truck ride before it arrives at your grocery store.

So whether it's because your food budget is getting squeezed as the costs rise or because you think food costs will increase aggressively as we run out of the easy-to-find fossil fuels, growing your own food is a great idea. From a hard landing point of view the decision to grow some of your own food is simply not an option. You have to. This will be one of your sources of sustenance. And like so many of the other new activities you'll be engaged in, you'll find it incredibly rewarding as well. There is nothing like eating food that has come from the seeds that you planted in the ground. It's something that's uniquely human and something that goes back to the time when our species stopped being hunters and gatherers and started to use agriculture as a means of self-support. You're going to really enjoy "The 100-Foot Diet" as you bring your homegrown vegetables into your kitchen.

I have been gardening for more than 35 years. When I was 16 I planted my first vegetable garden in the concrete-like clay of our suburban backyard. Today I run a CSA and supply 40 families with a box of produce each week for 16 weeks during the summer.

For those 35 years I've been reading books about gardening (I also wrote one in 2010 *"The All You Can Eat Gardening Handbook"*) and talking to lots of experienced gardeners. From all of this I've learned that growing food is really easy. You'll have failures and you'll have successes, but if you keep at it long enough you'll figure it out. You need compost to build your soil, water to hydrate your plants, and time to weed and nurture your vegetables. That's all. It's not rocket science. All you really need is motivation and you'll have a great garden. I hope my discussion on the challenges we face has motivated you enough to plant a garden. Now I'll share a few of the things I've learned over the years.

Getting Started

So where do you start? First you need some land. How much land? Well that depends. To begin, why not get started small with the intention of expanding. If you own a house, find a spot in the yard that you can spare. If you can't find any room to spare, something's got to go. Maybe it's the hot tub, maybe it's the gazebo, but your priorities have just changed so it's time to get serious about food. If your pool takes up most of your backyard, it's time to dig up all the trees and shrubs so you can plant vegetables. The time to worry about your yard looking fashionable is over. If you still own a house you should be happy to have the opportunity to grow your own food.

If you live in an apartment start with some garden boxes. Hopefully your balcony has enough sun to get some tomatoes going. Another great option is to find a local garden plot that you can rent for a season. Many cities are turning vacant property into community plots so local people can grow some of their own food. These also make an excellent place to meet other people who may share your world view and to talk to and learn from more experienced gardeners. Living in a community is going to be a good thing during challenging times. A community garden allows you to get to know more people in your city and form a group. This always helps the isolation so many people feel in the city and will give you greater resources when things go wrong.

To find a local garden plot start with your local city hall and see if they can recommend a place. Put up a sign at the local recreation center, ask people you already know in the community, find a local environmental group as they'll often have a listing of these, use the Internet, do whatever you have to do to find one. The one thing you may want to check is whether soil tests were done prior to turning the location into a garden. The unfortunate thing about cities is that there may have been a commercial or industrial enterprise there previously, and the soil may contain lead or other chemicals it would be good to stay away from.

If you don't have room and can't find a local garden plot another option is to find a CSA (Community Supported Agriculture) farmer close enough to reasonably get there to work on the farm. This way you'll be learning and getting the experience as well sharing in the harvest. Can you take your bike on the local transit system and ride to the farm from there? Do you own a car and could you devote one whole day on the weekend to working there (and the next day to recovering?) Get creative, but get your hands in some soil somewhere soon!

Soil Prep

Many people begin with land that is covered in grass. There are lots of ways to turn that grass patch into a garden. The most direct way is by using a shovel and turning over clumps of grass to expose the dirt underneath. If you're thinking about starting a garden next spring, and it's fall now, just turning over the grass should work fine. If you're a few months away from planting and there's no snow on the ground, lay down some old newspapers or spread out the cardboard that your new energy-efficient fridge came in and this will kill the grass over time. You might be concerned about contaminants in the newsprint ink, but many printers have switched to vegetable-based inks. (Also, since I don't eat organically all of the time, I'm not too concerned.) After a few months under papers or cardboard or even an old large pool cover you pulled out of the neighbor's garbage, the grass will be much easier to remove. If it's really dead you can use a hoe to pull it apart, but a shovel may still be required.

It's important to make sure you keep the soil that the grass is rooted in. That will be your best topsoil, and if you pull out the top four or five inches of grass and soil and discard them you'll be losing it. The topsoil will have the most potential nutrition for your vegetables and will be full of micronutrients and microscopic organisms that make soil healthy. Those organisms are going to break down compost and make the essential elements that plants need to access easily. If you have a heavy soil or healthy grass with a large root system that is taking a lot of topsoil with it, do your best to knock the soil off. Then pile the clumps of grass nearby so you can compost them and ultimately get all the valuable materials back over time. In much of my garden I turn over the grass in clumps and then use a multi-tined hoe and rake the clumps across an area that I've already prepared. Each time I grab the sod more soil falls out and eventually I toss the remaining grass into piles to compost.

If the area you're to garden was cleared before or has never had grass on it, you may be able to use a rototiller. This is a gas-burning machine and you may be loathe to use such a thing, but when you are just getting started it can save an amazing amount of work. I use a rototiller in my workshops as an example of the energy in a gallon of gas. Three tablespoons of crude oil represent the equivalent of 8 hours of human labor, so a tank of gas in a rototiller can replace a whole day of shovel and hoe work. I'm not recommending it; I'm just throwing it out there as a possibility. I wouldn't suggest you purchase a rototiller, but if you can find a used one or rent or borrow one in the spring when you need it, it can improve how much you can accomplish in a short period of time. If you have the time, then grab that shovel. I've already suggested that you scrap the health club membership, so this is one of the activities that will help you keep in shape. And long-term, as the price of gas increases and it becomes more and more scarce, it may not be possible to run a rototiller. If you've got some jerry cans of gas stored from last year, a rototiller is a good small engine to burn it up in so you can replace it with fresher stuff.

If you decide to use the shovel, purchase a file and sharpen the shovel blade. Most new shovels won't be sharpened, and if you're using a dull shovel blade it's going to make your job much harder. Also make sure you have a good-quality pair of workboots. Pounding down on the shovel with running shoes to push through densely matted grass is really hard on your feet. You need a boot with a very rigid sole, and if you're as dangerous with a shovel as I am a steel-toed workboot is recommended.

Rototilling over a patch of light grass or other plant material will break it up, but the grass will still be in the garden, so when you've finished use your cultivator or rake to remove the grass. This will keep it from taking root again and reducing your vegetable yield. If it's a large area of stubborn grass, rototill up and down the garden, rake and remove some grass, then rototill the garden side to side and rake again. It's really important that you try and start your garden with mostly soil and not a lot of roots and weeds that are going to have to be removed later. If you have access to a large amount of mulch like rotten hay or leaves from last fall you won't need the soil quite so pristine, since the mulch will keep the weeds down around your plants once you apply it. Weeds are going to be one of your biggest enemies as the summer wears on, so try and eliminate as many as you can right up front.

With your soil cleared the next thing you do is add compost. Compost is the key to a successful garden. If there is one thing you should get out of this gardening chapter it's the importance of compost. Compost, compost, COMPOST! I'm really serious. In all the books I've read and research I've done and market gardeners I've spoken to, compost is King. Compost is going to give your soil life. It's going to make it rich and provide plants with the nutrition they need for vigorous growth. It's also going to provide humus and the organic material it needs to help the soil retain moisture. The remains of grass clippings and leaves and hay and vegetable peels are all going to be in various states of decomposition and they will absorb water when it rains and release it back to the roots during dry spells. Good compost is the key to a healthy garden and good yield. There have been many times when I've spoken to lifelong gardeners about specific problems I may be having in my garden and invariably the solution is "more compost." They may just be saying this because they don't have any other pat answers, but it's always worked for me.

It doesn't matter exactly what is in your compost but you'll want a lot of it. It's great if you're composting all your kitchen scraps but these generally aren't enough. You need to supplement, and since you're growing a vegetable garden you need to supplement in a big way. Start by looking at all the natural organic material around you that ends up in a landfill. Some cities now have separate wet and dry pickup, but many haven't taken this step. The best you might get is leaf pickup in the fall, but this is a great place to start. Twenty years ago I lived in a city that had a crisis as its landfill was filling up, yet it didn't pick up people's leaves separately. In the downtown area where I lived, every fall there were thousands of green garbage bags full of leaves out at the curb to be trucked to prematurely fill up the dump. You could easily recognize them because a bag of garbage would have lots of pointy boxes and irregular shapes sticking out of it. A bag of leaves would be perfectly round and cylindrical with the indentations of leaves in the plastic. So on the night before garbage day I would set out with my wheelbarrow and load up 3 or 4 bags at a time and walk them back to my place. By the end of the season I would usually have about 100 bags of leaves in my pile behind the garage. My daughters loved it. By the next fall, after a summer of turning the pile over every two weeks with a hay fork, I would have a huge pile of wonderful dark, rich compost. The bottom layers would be full of worms that had come up through the soil to work their magic in the leaves, eating their way through and leaving dark rich castings to become part of the compost mix.

Eventually the city started picking up the leaves and asked residents to put leaves in clear plastic bags. This made my job even easier. Now many cities have asked homeowners to put organics like leaves in large paper bags that will compost down. Your neighbors are getting rid of a gold mine of organic material that you need for your garden. Ask them for it. I'll bet lots will even haul it over to your place as soon as they've finished raking just to get rid of it. Grass clippings are something you should be getting as well. With grass clippings, though, you might have an issue with pesticides. I always watched to see which neighbors sprayed their lawns and ended up with the little white warning signs. When I was grabbing grass on my garbage-night wheelbarrow runs I'd try and avoid those places. I even convinced a few neighbors to stop spraying so I could use their grass clippings, and they didn't seem to mind. Grass clippings are excellent for your compost. They are very high in nitrogen, which is one of the key building blocks of healthy soil.

Then it's just a matter of finding as many sources of diverse organic materials as you can. Is that local coffee shop throwing out coffee grounds? Why not provide them with some buckets that you'll pick up regularly. Perhaps that fruit market or even the grocery store where you shop has to dispose of fruit and vegetables that are no longer saleable. What about that local greenhouse/craft store that had the excellent fall display with all the corn stalks? Ask for them, since they may be going to throw them out. Or that church with the huge nativity scene with all those straw bales. "Can I have them when you you're through?" Crushed eggshells are a good source of calcium and your cabbages will like them, so ask the local diner with the busy breakfast trade if they'll toss their eggshells in a bucket for you to pick up after every busy weekend.

Get creative, but make enlarging your compost pile a priority. Once you start assembling these various materials you should start mixing them together. If you have many bags of leaves and grass clippings, hold some back. Take as large an area as you can find to make your compost pile. The days of your nice, neat little black plastic compost barrel are over. If you want to grow a lot of vegetables go big on your compost pile or go home. There are lots of designs for larger compost bins, but an easy one is wooden pallets. Stake them so that you have two sides and a back. As you build your pile some will fall through the spaces between the boards of the pallets but because they're open it allows air to get in. If you find two more pallets, make a matching compost area beside it and you can have two piles at different rates of decomposition, one that's garden ready and one that may not be ready for a few months.

Your compost should consist of layers from as many sources as possible. A layer from your kitchen composter, then six inches of leaves, then six inches of grass clippings, then an inch of coffee grounds, half a bucket of eggshells, then start over again. If you're in a rural area hopefully you can find some rotten hay and straw to include. Start mixing this in.

Every week or two, do some stretching exercises, then get your hay fork (with the long pointy tines, like the ones the villagers grab when they're going to chase monsters in the movies) and move the compost pile over three feet. You'll have to grab layers and sections at a time and heave them into the new pile. This will mix the pile up and get it decomposing faster. It would be good to have a garden hose handy and wet both piles as you work. The moisture will help speed things along and make a better environment for the organisms you want to attract to break down the materials. For the first month or so the pile won't look that different, but as the summer progresses you'll start noticing that more and more of the pile looks like soil. After eight or ten weeks of this you'll be amazed at how much it has decomposed and looks like those wonderful pictures you see on the side of the commercial composters with people putting in kitchen scraps at the top and beautiful black topsoil coming out the bottom. That has never been my experience with my big plastic composter. I find the materials sit and smell and turn into a big disgusting gooey mess. One of the main reasons for this is that the material doesn't get enough air.

Anaerobic decomposition refers to the breakdown without oxygen, while aerobic occurs with oxygen. So if your compost isn't getting any air it will not break down very quickly and it will smell. If you're turning it over every couple of weeks it will be getting lots of air, and the fact that it's now a pile means air can get into it and keep any smells to a minimum. You'll know your compost pile is working well if it's getting hot. That heat shows you the microorganisms are working hard because heat is one of the byproducts of the aerobic decomposition process. So the more heat the better. If you're turning the pile over on a cool day in the fall and you see steam coming out of it, you've hit the compost pile big time! "You've gotten to the rotten zone!" Celebrate "Decay Day"! If you blog, write a post and boast how you toast your compost. OK, I'll stop now.

I realize that a large compost pile is outside a lot of people's comfort zones. If you get too much grass in there and don't turn it often enough it'll get stinky. I'm just suggesting that if you're serious about having a good garden this is how you're going to do it. You can always buy commercial fertilizers, which will produce excellent results, but you're basically getting your soil addicted to fossil fuels. You'll need to keep adding this fertilizer because your soil will get lazy and so you'll have to keep using your after-tax dollars to keep buying something that adds much less long-term value to your soil than the compostable materials that your neighbors are throwing out, for free! If you do decide to buy things like bagged manure, keep your eye on all the garden centers, especially the temporary ones set up in grocery store parking lots. They'll often get pallets of these bagged products and by August will be ready to close up shop. The bags that are left may be beaten up, but this is when storeowners offer them at a heavily discounted price, and those bags will give you a really excellent return on your investment of time. Take your duct tape and offer the retailer 10¢ on the dollar to take the broken bags. If you can borrow a neighbor's truck, maybe you can grab the pallets to build your composter with at the same time that you're grabbing the bags of manure. The storeowners may be happy to get rid of it all in one fell swoop. They're going to have to move it when the season winds down anyway.

In my garden I'm looking to provide nitrogen and carbon. Grass clippings or any fresh green material, kitchen scraps, and manure are all high in nitrogen. Corn stalks, leaves, straw, and hay are good sources of carbon that also have some nitrogen as well as phosphorous and potassium that your plants need. When you buy a commercial fertilizer it will list these three items in order: N = Nitrogen, P = Phosphorous and K = Potassium (potash), or NPK. Typically a lawn fertilizer will be high in nitrogen to make your grass green so the ratio will be 21-7-7. A commercial garden fertilizer may be 6-10-4. The more you read about what combination of these is optimal for a garden the more intimidated you can become.

Obviously some plants will like more of one element and less of another. Some will require trace elements difficult to obtain from any given source. My approach to this has always been the shotgun approach. I spread as wide a path as I can in terms of what I put into my compost and garden and hope that I get what I need. This is not completely scientific I know, but there are so many variables with gardening and you can use up so much time planting and weeding and watering that sometimes I believe you can over-think this. If you can afford to purchase a commercial fertilizer and you've just moved into a new subdivision where all the topsoil has been stripped away, go for it. If you don't have access to a nearby horse farm for manure, buy some commercial bags of it. If you get lots of grass clippings in your neighborhood but not too many leaves because the trees aren't mature enough, make sure you don't overdo it. And when you have to be in one of the older sections of town in the fall, make sure you have an old blanket or tarp in the back seat so you can load up on bags of leaves at the curb. Try to get as big a variety of compostable materials as you can and you should be fine.

When we lived in an apartment we had a vermi-composter, a large wooden box lined with plastic that had "red wiggler" worms in it. We put our kitchen scraps in it and the worms did a fabulous job of breaking them down. Periodically we'd take the compost to my father-in-law's garden. When we moved into our house we put the red wigglers into the area where our compost pile would be and they thrived. Years later as I turned over the compost there would be masses of worms in it, breaking it down. There were handfuls of them! Even though our area experiences several months of below freezing temperatures, with so many bags of leaves piled in there in the winter the worms were all nicely insulated and back to work in the spring.

By the fall you should be able to start putting some of that compost on the garden you've been working on and turning over all summer. In the following spring take out some more and add it to the garden. Make sure you keep some of those worms you'll see in the compost piles and they don't all end up in your garden. It's great to have them in the garden, but on the scale of your garden they're best kept in the compost pile.

Manure is also an excellent soil conditioner. If you're starting a garden and you're in an urban environment and can afford commercial manure, go for it. Sheep and cow manure will come in plastic bags and will already have been composted down nicely for you. This is an excellent way to start building up your topsoil, which is the critical part of your garden. If you're in the country try and find a local farmer who has extra manure or a nearby horse owner who may have some to spare. You have to be careful with manure, especially cattle manure, because it may contain E, coli and other nasty organisms that can make you sick. If you can get manure find out how long it's been sitting. If you're unsure, pile it away from your garden for six months until you're sure it's safe. As with your compost pile, make sure it gets hot, as the heat kills pathogens that you don't want near your food.

I realize this section on compost is long and drawn out! Yes, it is. But there is a point to it. If you don't start with well-conditioned soil, everything else is a huge waste of time. You may enjoy some success for a while but the lack of soil conditioning will catch up with you and you'll never know whether those black spots on the leaves are a blight or just a plant in distress because its roots couldn't find the nutrition it needed in the soil.

Planting

One of the greatest things about gardening is the anticipation of spring, and one of the best ways to hurry spring along is to start some of your own plants indoors late in the winter. By March your local greenhouse will be in full swing planting flats of flowers and vegetables for customers to buy in April and May (depending on where you live). To save yourself some money, you should be doing it yourself. You'll get a better selection of seed types than the vegetable plants from your local retailer, which is one of the other advantages of starting early. Many people swear by their "heritage" tomatoes and claim that they taste infinitely better than standard commercial brands. Some people have started saving and swapping seeds with others to increase their variety. And there is nothing nicer on a cold winter day than sitting down with a seed catalog and going through the color photos of all those fruits and vegetables that you'll be able to grow next summer.

Waterproof seed trays are an excellent way to start seeds, especially if like us you'll be moving them on a regular basis. Some people go to the expense of purchasing grow lights because it's often hard to get enough sunlight in your home in the winter months for seedlings to thrive. These are a great idea. Living off the electricity grid we have yet to invest in them. We start our seeds in seed trays with a plastic cover, like a small greenhouse, and we move them from window to window during the day as the sun moves. The greenhouse lid keeps them warmer at night. Eventually as the days get warmer we can put the whole tray outside so that the plants can get full sunlight. This is called "hardening off" the plants so that it's not such a shock when you do finally put them into the garden. It may sound like a lot of work moving them to follow the sun during the day, but it also keeps you watching to see what's coming up and how the seedlings are doing.

Starting seeds indoors will be something you'll have to get a feel for based on what you like to eat, when the last frost hits your area, and how big your garden is. We've developed a hybrid system where we start some things ourselves and buy others from a local greenhouse. Some

Our homemade shelves with growlights to start plants in winter.

days we don't have the electricity to have grow lights going 24/7 so we know that things like tomatoes and pepper plants will not thrive and will be gangly plants that are too tall and weak. By the time we are ready to plant tomatoes in our garden, the commercial greenhouse has tomato plants that are thick and lush and compact. They might cost us $1.49 for four plants, but they are beautiful plants that thrive, so this is a tradeoff we're prepared to make. We do start some of everything and we do put some of our own tomatoes in the garden but we supplement them with greenhouse tomatoes to make sure we have a good mix. We generally start our plants in March and then start another set in April and more in May. We do this so that we have plants at various stages to put into the garden. The challenge with purchasing commercial plants is that if you plant four broccoli and four cauliflower when you put the garden in after the last frost, when they mature you'll suddenly have more broccoli and cauliflower than you can enjoy. Some of it will end up in the freezer but we make sure than we have a number of plantings and have everything at various stages of maturity to extend the season.

Once the ground is warm enough to be worked you'll want to get out there and get the garden ready. Remove any plants that you left in from last year and weeds that didn't get pulled in the fall. Work some compost in and rake the soil to get it ready. Some things can be planted early, like spinach and radishes. We'll often have seeds self-germinate from last year and it's amazing how early they'll germinate and begin to grow. Sometimes there's still snow in certain parts of the garden and spinach plants will have started growing. Lettuce seeds are easy to save so I usually put in a row of these early. Peas can also tolerate cold weather and in fact I find they like a frost. It's always nice to get these early vegetables going to entice you into the garden after a long winter. There's nothing like that first salad where you use some of that early lettuce and spinach to remind you that spring is here.

There is much debate about how much room to leave between plants and rows. This is almost a bit of a city/country debate. There are many books about bio-intensive or "square foot" gardening that suggest you can grow huge amounts of vegetables in a very confined space. The books that discuss this are fairly technical in their description of how to do this successfully. If you have limited space I would urge you to learn more about this, but with the space I have it's not a technique that I require and so I am not going to go into it in great detail. I would suggest you start by following the seed package recommendations on how far apart to plant things. If you have limited space and the seed package recommends rows 3 feet apart, try making them 2 feet apart or even just 18". By harvest time the plants will have crowded out the walkways between rows and it will be harder to get around the garden. But by then your priority will be harvesting and eating those vegetables, so if the garden looks a little unsightly it will be a small price to pay. If your vegetables are this dense it also makes it harder for weeds to grow, so there is some benefit to having tightly grown plants.

My attitude about bio-intensive gardening is simply that plants need nutrition and water and there is only going to be so much of that in any given area of the garden. If you crowd too many plants into an area that can't support them you'll simply reduce the harvest. So you really have to be paying attention to do this well. If you were starting a garden without a good handle on the quality of your soil, I would recommend that you go with rows a reasonable distance apart. As you get more comfortable with your garden and you've been able to provide more compost and supplements you can begin increasing the density of your plants. After a few seasons you'll also have enough experience to be able to decide how much of everything to grow. After a summer of realizing that no one in the family likes kohlrabi and that a 12-foot row of beans grows more beans than you'd eat in two years, you'll be better prepared to scope out a strategy for intensive growing.

You should always rotate where things grow in your garden. This ensures that insects and diseases that remain in the soil are less likely to have an easy time getting to your crops. Each year I change the orientation of my garden with rows going vertically one season and horizontally the next. I make sure that my heavy feeders like corn are moved around each year. This also allows me to ensure that the walking paths are different each year so I don't end up with severely compacted soil.

One thing you should use to extend your season and save money is a cold frame. This is a small greenhouse you use to help start seeds in the spring and keep some sensitive crops going in the fall. There are lots of plans available but I built mine out of four windows I found at the local dump. They were old and have a wooden frame so I just screwed three of them together and put the top one on a hinge to open and close. I left the back open and have it against the concrete foundation of my house. This way the sun warms the concrete during the day and the heat is radiated back at night. A cold frame will keep plants from getting nipped by frost. It's also a great place to put your flats of seedlings after the sun goes down. When it's warm enough you can move them outside during the day to harden off and back into the cold frame at night to protect them. We also start some lettuce and spinach in our cold frame in the fall and it gives us salads until December. Growing in a colder climate as we do we try everything we can to extend our season. So keep your eye on your neighbor's garbage and always grab any old windows you see. If you have a few of them you can make a cold frame and then if you accumulate enough you may be able to make your own greenhouse someday. Since your garden is reducing the distance your food travels you are actually increasing your greenhouse glasses to reduce your greenhouse gases (loud groan)! I would like to thank my brother-in-law for that pun. If you disapprove contact him directly.

Your local weather or government agricultural office can provide you with a final frost date if you're not familiar with it. As you get closer to that date you may want to start putting in some cool-season vegetables like beets, broccoli, cauliflower, cabbage, and beans. If I'm going to make my rows ten feet long, I'll seed half of it a week or two before the last frost date and the other half a week or two later. Hopefully some of the seeds will germinate and will get a bit of a head start on the second planting. Some seeds need the soil to be fairly warm to germinate so I'll wait until our final frost date to plant carrots, celery, potatoes, and onions.

That frost-free weekend will tend to be my big gardening weekend when I'll plant as much as I can. I'll plant corn as well as my "vine" plants like squash, cucumbers, muskmelons or cantaloupes, and pumpkins. You have to be sure that the danger of frost has passed and the soil temperature is starting to get high enough for the seeds to germinate. This is also the weekend I'll pick up the bulk of my vegetable plants from the local greenhouse. These will be my real heat-loving plants like peppers, tomatoes, eggplants, and watermelon. I'm fortunate that I have an old concrete barn foundation on my property. This is a real advantage for these heat-loving plants in a colder climate like mine. Even though the danger of frost has passed, night temperatures can still be quite cool for a few more weeks and this slows down the growth of these plants. I built raised beds in the barn foundation because it has a concrete floor. I keep the soil in place with old cedar posts that I retrieved when the road crew replaced some of the guardrail posts along our road. They are just cedar trees and are not chemically treated. The beauty of the barn foundation is the thermal mass of the concrete. When the sun is out during the day the concrete absorbs heat, and once the sun goes down that heat is radiated back. The concrete walls also act as windbreak for some of those cool breezes and help nurture these fussy plants that act as if they need a Caribbean holiday to get growing.

Be careful with real heat-loving plants like peppers and eggplant. While I buy them with everything else on May 24th, I don't actually put them in until the middle of June. I find that if they're in the garden and the nights are too cool their flowers won't set fruit properly later on. I keep them outside during the day but keep them inside at night until I'm ready to plant.

Raised beds are a fine idea if you have a limited amount of space. They allow you to grow fairly intensively without compacting the soil. The roots of your plants need oxygen and if the area around them is heavily compacted they may have trouble accessing it. With a raised bed you do all your walking around the outside of the bed, which leaves the soil fluffy and aerated. Since my garden ends up with well-worn walking paths, I like using my rototiller to loosen that soil up in the fall and again in the spring before I plant to ensure that the roots have an easy time making their way through the soil in search of water, nutrients, and oxygen. With a raised bed be really careful with the material you use for the sides. Use rock or concrete block if you can. Cedar or other untreated woods are nice but may be expensive unless you can reclaim some. Do not use railroad ties or pressure-treated wood. Railroad ties are treated with creosote, which is a very nasty coal-tar-based polycyclic aromatic hydrocarbon, which you don't want near your food. Likewise some pressure-treated wood can contain copper arsenic, which is a nasty chemical that you don't want near anything you eat. With the concern about the safety of this chemical in residential applications many companies are no longer treating their lumber with it, but I would still recommend you stay clear of pressure-treated wood. Try and make the edges as natural a material as possible.

One of the advantages of a raised-bed garden is that the soil will get warmer faster because more of it is exposed to the sun. So you may want to have a raised bed to get your heat-loving plants going early. The downside of a raised bed is that it will lose moisture more quickly, both to evaporation and to excess water running out the sides during major downpours. So if you do plan on using raised beds you'll have to be more vigilant about watering them. If you're invited to a neighbor's cottage for a weekend in the hottest, driest part of the summer, make sure you find someone to water daily; otherwise you'll be very disappointed when you get back.

We love tomatoes so I plant a lot more than we need. I purchase plants from the greenhouse and start some as well, just to add variety. I try to make sure that I grow several different types of each vegetable I plant. This improves the chances that with varying conditions at least one variety will produce well. Since the greenhouse plants are invariably bigger and healthier than the ones we start indoors, they ensure that I get stuff to eat as soon as possible. If some start ripening much later that's fine; those are the ones that we'll can or freeze or pass along to our neighbors, especially the neighbor who keeps us supplied with horse manure.

I also use this technique with other plants like peppers, broccoli, cauliflower, vines like squash and watermelon, and eggplant. We start each of these indoor late in the winter as well as several times as the spring progresses, and we also buy commercial plants. This means that I'm varying the time that things will ripen so that I can try and have as steady a supply of produce as I can during the growing season.

Let's use broccoli as an example. While some well-known public figures aren't big fans of broccoli, it's an amazingly healthy food noted for its cancer-fighting properties, so we like to have a supply available from the garden all summer. So on our last frost day we put in four plants from the greenhouse. I also plant half a row of seeds that weekend and put some of our own plants, which are two or three weeks behind the commercially grown plants, in the remainder of that row. About two weeks later I'll put in another half row of plants that I started. You can do this until July. Broccoli is one of those plants that likes cool weather and can handle a frost. Invariably, despite my best strategies, there'll be a few weeks in the summer when we have way too much broccoli, so those are the weeks we freeze a few bags and give some to the neighbors. We'll have broccoli into September and even October here, even after we've had frost. By then I'm usually pretty sick of broccoli and will take a month or so off before we start eating the frozen stuff.

You should employ a similar strategy. You'll be tempted to get out there that first weekend and put everything in and fill up the garden completely. You'll only do that once, because come July you'll have more vegetables than you can eat. The next year you'll start planting as early as you can and putting new stuff in every week until July. This will spread the season out nicely.

I find that in some years plants start better as seeds and in other years they do better as transplants. I plant my vines like muskmelon, squash, watermelon, zucchini, and pumpkin in small hills with lots of room around them to grow. I usually plant a few transplants already well under way and some seeds as well. Some years the transplants will take over and thrive. Some years they don't but the seeds pick up the slack. I like my plants to have a "Plan B."

Pests

I also do this because of the issue I have with pests. There are many small critters out there that will be thrilled that you've provided a wonderful healthy diet for them, and they'll do everything they can to eat it all themselves. The biggest problem I have is with cutworms. Cutworms look like caterpillars and they hide in the soil over the winter. In the spring they wait for green growth to eat. They will come out at night and wrap themselves around your plants and cut them off just above the soil. Then they'll pull that plant down in to the soil and munch on it during the day. It sounds pretty cute, but a couple of cutworms can take out a whole row of peas or beans before you know it. I deal with them in a variety of ways. My first strategy is to plant more seeds than I need. Then as they sprout I keep my eye on the plants and look for the telltale signs of cutworm activity, either plant stocks sitting there with their tops missing or the top pulled into the ground. If you're not closely examining the rows on a regular basis cutworms are even easier to spot because you'll look at a row and you'll see areas with lots of seedlings started and then big holes in the rows. I just head right to those areas and start carefully digging through the soil until I spot the cutworms. I use the plural because they often work in groups. Sometimes I throw them in a margarine tub with water to drown them, but usually I'm so fed up with them trashing my beautiful seedlings that I take great pleasure in just squishing them with my fingers. I usually take a "live and let live" attitude towards creatures, but I draw the line at mosquitoes and cutworms.

To protect the transplants in the garden, I take a toilet paper roll, cut it lengthwise, and then cut it in half. I then wrap half a toilet paper roll around the bottom of the plant and push it down half an inch into the ground. This usually keeps the cutworms away. I also use a sacrificial plant strategy. Using seed from last year, I start a large flat of some plant that germinates easily. Lettuce usually works well. In between each of my peppers or tomatoes I put in the two- or three-inch-high lettuce plants. Over the next few days cutworms in the vicinity will be drawn to the lettuce, which they'll slice off. It's easy to see the lettuce knocked down and easy to retrieve the culprits responsible for the work.

My other major pests are potato bugs, or the Colorado potato beetle. These will overwinter in the vicinity of your garden and will wait until the potatoes start sending up shoots. Then they'll fly over, start munching on the leaves, and lay their eggs on the underside of the green growth. Potato bugs are fairly easy to spot, so I always start a row of potatoes very early, even before it would be normally considered safe. This row will attract all the potato bugs in the area and make them easier to spot. Then for the next few weeks I'll keep checking this row daily and drowning any bugs I find and scraping any eggs off the underside of the leaves with my fingers. These eggs clusters are easy to spot when you pull back the leaves while standing over the plant, because they are bright orange against the green of the leaves. By planting one row early you'll find that the bulk of the potato bugs will be drawn there and you can concentrate on eliminating them. As you start other rows later you will have some bugs but they will be far fewer. By then your garden will be in full swing so you may miss some, but a plant can handle a few bugs. If one of those egg clusters does hatch in a week or two your potato plant will be full of rapidly growing orange potato bugs. They are easy to spot and you can use a badminton racket to knock them into a bucket that has an inch of water in the bottom to drown them. Stay on top of them though because if you let them they will eat all the leaves that are taking the energy from the sun and storing it in the potatoes underground.

I focus on potato bugs because I think you should put a lot of effort into your potatoes. This goes back to the food chapter, where I suggest that potatoes are the perfect food to grow in your garden, rich in energy, protein, and vitamin C. They store well, so you can eat them all winter, and the ones that you don't eat you can plant next spring and start the process all over again. Planting one potato will net you eight to ten potatoes in the fall depending on the variety you grow and what sort of summer you have. Once you buy seed potatoes you'll really never have to buy them again. Every few years I buy a few new varieties just to keep some different DNA coming into the gene pool. For several years I got too reliant on Red Pontiac potatoes that we really liked but that were susceptible to blight. Some of the potatoes ended up with a dime-sized black spot in the middle called hollow heart. This wasn't a problem if you were cutting them up to boil or make home fries, but if you were baking them you wouldn't spot this until after you cut it open. So now I make sure we have at least three varieties. A seed catalog will list the characteristic of each variety. You'll want to get some that mature quickly so you can eat fresh-dug potatoes as soon as possible, and you'll also want some that mature more slowly and are better for storing.

For 30 years my garden has been a great experiment in seeing what works and what doesn't. I get a large sheet of stiff cardboard and I tape a large piece of paper to it and draw a map of the garden. I attach a pencil to it with a wire so I don't lose it. Then every time I plant things I note the variety. This row was "Lincoln" shell peas and this row was "Sugar Sprint" edible pod or snow peas. Then next winter when I'm sitting down to order my seeds I decide if I want to try that variety again. You can also keep your seed orders each year in a binder, but I find keeping track of things as I plant them helps me remember names. Eventually snow peas will form a pea and look like a regular pea pod, so if you're doing a stir fry and want to make sure you get the edible-pod peas, a map of what you planted is a great aid.

As the summer progresses I continue to plant new rows of seeds to keep a steady supply of vegetables ripening. Beans will keep producing for three or four weeks but eventually many of the beans will be too large and tough to enjoy, so it's nice to have some new plants that will provide you with more tender beans. As for the beans that get too big to cook up with dinner, you can leave them until the fall. At that point once the pods dry out you can pick them and dry them completely. Then during the winter as you're sitting watching TV you can open up the dried pods and remove the bean seeds. You've got a couple of options with the seeds. These beans are just like the ones that come in a can of baked beans or chili, so you can use them in cooking. You also can save to plant next year. Remember, it's always a good idea to put in more seeds than you need and thin the rows if too many germinate.

Since you're going to plant more seeds than you need, you're going to have to thin those plants as they emerge. If you look on a seed package for beans it may say that plants should be spaced 12" apart. If you saved seeds from last year and sowed them very densely you may have 10 or 20 come up in that 12" zone. It's always hard to visualize what your garden will look like in August as you see those plants emerge in May, but you have to try to because most of these plants will get very big, especially if you've been aggressive in your compost application. So ideally you won't want too many plants too close together because they'll crowd each other out. They'll compete for water and nutrients and this added competition will reduce the yield of each plant. So you are actually better to try and follow the directions on the package. I say this as the worst plant thinner on the planet. Oh, I've got my reasons not to, namely those cutworms that I never really trust to stop lopping off my seedlings. At a certain point your plants will just be too big and strong for a cutworm to gnaw through, so you have to thin. It's the hardest thing you'll have to do in the garden because it goes against every one of your plant nurturing instincts, but it has to be done. Sometimes I do it over several weeks, which makes it easier and gives me some reassurance that I won't thin to a reasonable level and then have cutworms take out the few plants I have left.

In terms of carrots I do some thinning but leave them pretty dense. I find carrots can be very finicky to germinate, so once they do germinate I hate to thin them too much. Carrots will also grow fairly compactly if you leave too many in one spot. When I thin I make sure that I wet the soil well first, which makes it easier to extract the plant you want without damaging neighboring roots too much. Since carrots are slow to germinate, by the time I'm ready to thin them they usually have a pretty good crop of weeds mixed in with them too. In fact, my thinning is probably better described as a good weeding that takes some carrot seedlings as I go.

Seed Saving

Pea plants are like beans, so if you don't get a chance to harvest them in time and the peas get too tough to enjoy you can just leave them until the pods go brown and harvest the peas to use as seeds for next year. Lettuce and spinach plants that are allowed to grow without being harvested will eventually send up a seed shoot, which will produce wonderful flowers. If you leave these long enough those flowers will brown and you'll end up with dozens of small seedpods. You can harvest these seeds and use them next year. When you cut your first broccoli the plant will continue to send out smaller shoots that will be smaller heads that you can eat as well. If you don't harvest all of these, the broccoli head will become a mass of flowers that will form seeds that you can harvest for next year once they dry.

Plants like these produce seeds every year (annually). There are other plants that produce seeds every second year (biennially). Carrots and cabbage have to grow one year and then be overwintered and planted again the following year. That second year they send up shoots that form flowers and then seeds. Depending on where you live and how cold your winters are, you can sometimes leave these vegetables in the ground over the winter and they'll start their seed cycle the following year. If you get very cold winters you can either heavily mulch them with straw, hay or leaves or you can pull them up with some root on (in the case of the cabbage), put them in a bucket of sand in your root cellar, and then plant them back in the soil the following spring. As you spend more time gardening you'll get better at seed saving; it is a financial boon, and it will also be a good way to meet others who may want to swap you for the seeds they've saved. It's just like trading baseball cards, but these trades will provide you with food!

Some seeds will germinate in your garden without your help. Each spring you'll find lots of "volunteers," seeds that fell off vegetables that you let go to seed last year and are now coming up on their own. My policy usually is to get rid of volunteers or move them where I can use them as sacrificial plants to attract cutworms. Sometimes these seeds will come from plants like tomatoes. Any tomato that ends up being left on the ground will be filled with hundreds of seeds. Sometimes these volunteers come from plants that are "hybrids," having taken characteristics from several plants. If their seeds are allowed to germinate they don't always make the best plant if left to mature. Over the years I know I have had some hybrid carrot seeds end up in my seed collection, because if I have a thick, wide row full of carrots, every tenth carrot will be white and won't taste good. I just discard this and don't worry about it because the other nine are fine. These white ones may be coming from wild Queen Anne's Lace, which is a wildflower also called "Wild Carrot" that grows around the garden and can cross-pollinate with carrot flowers. There are several excellent books on seed saving and if you find yourself enjoying the satisfaction that comes from being able to generate your own seeds each year you should read one of these books.

Watering

After compost the other key to a successful garden is water. If you're growing in a city with town water you're lucky. Many parts of North America experience water shortages every summer, which can be very hard on a vegetable garden. So as I discuss in the water chapter, you need to have as many rain barrels as you can, especially if you're serious about growing a good portion of your own food. On a smaller scale you'll be able to use watering cans to water your vegetables from your rain barrels. Try and water in the morning because it will allow the water to get to the roots without rapidly evaporating, which it will do later in the day once the sun comes out and warms the soil. If you have a large garden and limited water you may want to use mulch to help with water issues.

Mulching materials are basically the same things that you're adding to your compost heap; but instead you're just going to apply them directly to the garden. So once your plants are up you can put shredded leaves, grass clippings, straw or hay, or wood shavings around the base of each plant. If you're putting in transplants you can put mulch around right away. Some gardeners like to get mulch on as early as possible to try and warm the soil quickly and inspire vigorous growth early on. You'll want to water mulched vegetables heavily to make sure the water gets through the mulch and gives the roots a good soaking. Then when the sun comes out the mulch will keep that moisture from evaporating out too quickly.

Obviously if you're putting leaves or grass on your garden and keeping it wet it's going to break down just as it would in a compost pile, and this is a good thing. You're adding that organic material to mulch the soil at the same time as you're helping to preserve some of the moisture the plants need. You can appreciate that with a large garden you're going to need a lot of mulch and a lot of inputs into your compost heap. This is where you'll have the crisis of conscience each fall: "Do I take just one more trip 20 blocks over where I know they've got bags and bags of leaves out at the curb? I'll burn gas, but I need those leaves!" Perhaps it's time to contact the city and see if they'll dump a load of leaves on your driveway rather than driving them miles away to the municipal composting facility. You can expect the initial reaction to be, "We've never done that and can't help you," which is why you'll need to contact your municipal councilor and get her working on your behalf. Many municipalities will let you go to the site to collect the compost once it's been created, so why not suggest that you're just saving all that diesel-fuel expense by having leaves delivered to your place?

As our summers get hotter many cites are also issuing watering restrictions, which is just one more reason to make sure that when it rains you collect lots of water. If you're in the midst of a drought and you know a big storm is on the way it's time to get creative. Maybe that old kiddie pool that isn't used any more can be filled up once the rain barrels are full. Vegetable plants need water and they will go into distress if they don't get enough of it. One thing you want to make sure of is that you encourage deep root growth, and you do that by watering less often but very thoroughly. In other words, rather than watering every day and only wetting the top one inch of soil, you should water every third day but put multiple cans of water on so that water gets down five or six inches or more. The deeper the roots the less susceptible the plant will be to drought damage. While you'll feel better putting a bit of water on every day you're not doing the plant any favors. Tough love means watering every few days but watering vigorously.

As your garden gets larger you may want to look into an irrigation system. A drip irrigation system can be hooked up to a raised rain barrel, a water faucet on your home, or even a solar-panel-powered pump from a nearby pond. A pressure regulator ensures that a slow and steady amount of water is gradually applied to the plants out of holes in the pipes 12 to 18" apart. This follows my deep-water recommendation by putting water on gradually and letting it spread over a large area and soak down deeply. When you water like this the soil's capillary action draws the water both horizontally and vertically. Think of dipping the corner of a big fluffy towel in water. Even if you just dip it in for a few seconds and then remove it, if you keep doing this over time the entire towel will be soaked because of the capillary action of the water moving throughout the towel. In a sandy soil the capillary action will tend to be more vertical, drawing the water down more than out. In a clay soil you'll have the opposite effect, where the water will be drawn horizontally and not as deeply. Hopefully your soil is a mixture of these types and will move water both ways.

One way to ensure that water travels well in your soil is to limit compaction by keeping the soil around the plants well tilled and making sure that your soil has lots of humus or organic material. Where does your soil get that? Compost! It all comes back to compost. Keeping your plants watered and growing them in a rich well-composted soil are the two keys to a bountiful harvest.

As the summer wears on you'll get into the groove of watering and looking for pests and enjoying your bounty. So many of the things you grow are best eaten while standing in the garden. Those first peas that you take out of the pod, rich in iron and fiber and sweetness, are a real treat and provide the greatest health benefits when eaten raw. Most of our peas never used to make it to the dinner table, so now we grow even more so we can have some with meals and freeze others for the winter. Even beans taste better eaten raw. If you grow your carrots as densely as I do and your soil is moist enough you can thin them in the summer, eat what you thin, and give its neighbors room to grow.

I grow vegetables like onions in a variety of ways. Most of them I grow from bulbs that I get through my seed catalog or the local feed mill. I transplant some that I've bought from a greenhouse, and I always throw some seeds in as well. I also grow every kind I can find including red,

Spanish, and cooking onions. I grow onions close together to start, and then as they get bigger I thin every other plant to use right away. This leaves room for the remaining onions to get really big.

Garlic is a unique member of the Alliaceae family, which includes onions. It is planted in the fall. The bulbs overwinter and are one of the first things to send up shoots in the spring as soon as the snow is gone. They grow vigorously in the spring and early summer and are ready to harvest in July. You'll know it's time when the lower leaves start to brown. After you've harvested the garlic and hung it to dry, you'll have more room in the garden to start some late beans, lettuce, and spinach.

Each year you'll get better at selecting those vegetables that do well in your garden and do well in your kitchen. There is always a tendency to grow every vegetable you can imagine and then discover that you don't end up eating a lot of them. I love turnips, once a year, at Thanksgiving. So I don't worry about starting them very early and I don't plant too many. Every year we plant kale and Brussels sprouts. These vegetables are health titans and if eaten daily would make you the healthiest person on the planet. Unfortunately I find it difficult to eat too much of them. The good news is that they keep well into the winter and they stick up through the snow in our garden, so hungry deer also enjoy some of that healthful bounty in the dead of winter when they really need it. This is a luxury you probably don't have if your garden size is restricted, so start deciding what you ate and enjoyed last year and what didn't seem to be so popular and move your garden in that direction.

More and more I focus on vegetables that store and keep well. As I discussed in Chapter 15, potatoes are the foundation of both my diet and my garden. Next come carrots, which are so rich in beta-carotene and antioxidants that they are one of nature's superfoods. I put my harvested carrots into buckets and fill up the buckets with moist peat moss and they last all winter in the root cellar. When they start getting a little soft as we get closer to spring it's time for more carrot soup and carrot cake! Onions store extremely well and complement the carrots and potatoes, so I grow lots of them. I also grow a lot of garlic, which lowers your bad cholesterol and helps fight colds. These all store without any energy required.

Next come things we can freeze like tomatoes, beans, broccoli, cauliflower, and peas. These are great in soups and sauces all winter. We also grow a lot of basil, which is wonderful to add to recipes during the summer. To preserve it we chop it in a blender with olive oil and freeze it in a thin sandwich bag or ice cube tray. Then it's easy to cut a chunk off and throw it in with a soup, pizza sauce, or even scrambled eggs to give them that "upscale" look and taste. I call them "basil-infused scrambled eggs." They're incredibly pretentious and our country neighbors moan when I wax poetical about them.

We grow a lot of corn because we have the room and because I love corn. When you see the size of the corn stalks you know they take a lot out of the soil, so I make sure I move them around each year. If you have limited space it may not be the best vegetable to grow. It's best if you have 3 or 4 rows of at least 15 or 20 plants so that they will pollinate properly. When I grew corn in the city the squirrels seemed to get more corn than we did. Here in the country the raccoons stay away from our place until the corn is ready. That's when our dog sleeps in our fenced garden to keep the raccoons at bay. Commercial corn can have a fair amount of pesticide on it, but fortunately we've never had to spray and still don't have major pest problems. Once we start eating our corn we always cook way too much; then with what's left over we slice off the kernels and put them in the freezer. With our Thanksgiving and Winter Solstice feasts this year we had sweet wonderful corn from our garden. An important reason to give thanks.

As the summer winds its way into fall your garden should still be providing lots of food for your table, and hopefully the weeds will have slowed down as well so that less maintenance is required. This is the time of year when some of those slower-growing vegetables like squash will be ready to eat. Since they're going to end up in the root cellar, I harvest and store the best specimens and eat the others. Any sort of scabs or blemishes on vegetables will tend to provide an entry point for the microbes that will bring about early spoilage, so we try and store only the nicest looking ones.

Now it's just a waiting game to see how long we'll go before a frost. All those beans and things you started when the garlic was harvested in July are now at their prime and you'll find yourself listening to the weather forecasts to see if there's a danger of frost. When the warning finally comes you may want to decide if there are some vegetables you particularly like that you want to keep going. In the patch where my garlic was I've usually planted beans, lettuce, spinach and a few broccoli and cauliflower plants, so I throw a big plastic tarp over it to protect it from the frost. It's important to remember to take it off the next morning because if it's sunny it can get pretty hot under the tarp. I continue fighting nature for a week or two until I finally just resign myself to the fact that nature bats last and just let it take its natural course. Some plants will survive and some won't. Before we let that last big frost hit we make sure the freezer is full of beans and tomatoes and other things we want to freeze. Some plants like broccoli, kale, Brussel's spouts, and squash will all do fine and won't be damaged by an early frost. In fact some will actually start tasting better as the cold weather enhances the natural sugars in the plant. I still may not have stored all my carrots and potatoes yet but that's fine, because they're underground and even though their tops might get nipped the root that I'm concerned with will do fine.

It's your choice whether you're going to leave the garden until the spring or start cleaning it up in the fall. If you've had any problems with tomatoes or peppers remove those plants and don't put them in the compost. If there's blight or disease it may just get back in the soil for future years. I tend to get distracted by the fall and leave much of the organic matter where it is. I'm afraid that by rototilling it in the fall I'll just leave exposed soil which is more likely to be blown away by fall and winter winds and washed away by spring rains. The one thing I do have to do is clear a section and plant my garlic. I try and do this before it gets too cold, just because it gets hard on your hands if it's too frosty out. I've always been concerned about putting those garlic bulbs into the ground just before it freezes up, but every spring they shoot right up and thrive. They are the first bits of green I see in the garden every spring and they are a welcome sight.

Every year I try new things, experiment with new varieties, and plant things differently. This year I tried planting corn in amongst the squash vines as the people native to America have done for centuries, but I didn't have much luck with it. A few years ago I tried artichokes, but they didn't mature well so I didn't try them again. Next year we're going to try planting peanuts just because they look cool to try. Every season brings failures and successes and a root cellar full of healthy organic vegetables that displace a fair amount of our grocery budget. Meanwhile we're displacing thousands of pounds of carbon dioxide that would have been generated trucking this food to us. And when the ice storm hits and we know we can't get to town for a few days, it's no problem. Our garden is still providing us with the sustenance of life. Now put down this book, pick up your shovel, and start turning over that grass in your backyard. There's a grocery store produce section out there; you just need to visualize it and make it happen!

17 Water

Water is really important, yet it's amazing how many songs include lyrics about stopping the rain. Sure it can be an inconvenience, but as you get older and start to realize just how important it is, rain really is something to be celebrated.

Up to 60% of the human body is water. The brain is composed of 70% water, the lungs are nearly 90% water, and about 83% of our blood is water. Humans must replace a little more than half a gallon (2.4 liters) of water a day, some through drinking and the rest from the foods eaten.

Water is becoming as big an issue as food and energy throughout the world. Wars are being fought over it. North Americans can be pretty wasteful with water. It's a precious resource and if you start treating it today like the precious commodity it is it'll be much easier to cope when it's not so conveniently available. So it's time you started getting in touch with water.

You're going to have 3 main water issues to deal with;

1) enough safe drinking water
2) water for sanitation such as washing and toilet flushing
3) water for growing your food

Some of the water you use you may be able to recycle. After you have a shower or bath, or wash vegetables or dishes, there's no reason you can't use that "gray water" for flushing toilets or watering gardens. As long as you use it for a "lower" purpose there's no reason not to be reusing water.

So let's look at each of your main water uses.

Drinking Water - Urban

Food-borne diseases and illnesses are prevalent throughout much of the world and are one of the reasons so many North Americans have problems when traveling abroad. We simply haven't built up the necessary immunities to giardia, schistosomiasis, and so many of the other common "bugs" that lurk in water in other countries. We know that health care will not be as easily accessible in the future as governments become increasingly challenged to provide services, so not getting sick from poor quality water is going to be more important than ever.

Cities provide this amazing resource—clean drinkable water—but less than 5% of it is actually used for cooking and drinking. Of the rest, 40% goes to flushing toilets, 35% to bathing, and 20% to laundry and dishes. Then of course there are the other frivolous uses like washing cars, watering the grass to keep it looking pretty, and washing driveways—the sort of activities that remind you of a 1960s episode of *Leave It to Beaver*.

As all levels of government become taxed, it's going to be harder and harder for them to maintain the level and quality of water they provide to city dwellers. Chlorine and some of the chemicals they've used in the past will become prohibitively expensive or in short supply. Underground pipes that need to be replaced periodically will not be, and the infrastructure that provides you with your water will become increasingly problematic. So regardless of where you live you're going to need to ensure a safe and secure source of drinking water. Many places in the world experience times of the day when they have no power and water, and while it seems like a remote possibility in North America it's worth being prepared for. This means having a supply of drinking water on hand whether you live in a house or an apartment.

How much drinking water you store will depend on how many are in your family, but a person requires at a minimum about half a gallon of water a day. FEMA suggests a gallon per day for drinking and sanitation. You'll probably need more if you are in a warm climate. I'd pick at least a gallon a day per person or more depending on how long I think I'm going to go without access to it.

Governments are now encouraging people to stock up on water in case there are weather-related emergencies, so having water on hand is becoming commonplace. Every newscast showing preparations for a hurricane shows people stocking up on water and being distraught when stores sell out. Get it now.

Don't forget that if you own a home you already have a fairly large water storage source available in your hot water heater. These can be up to 60 gallons. I'm assuming that if you're in a situation where you need this water, it is likely that the natural gas or electricity will have been off for a few days so the water in the tank will have cooled off. Hot water tanks have a drain at the bottom that some people, especially in rural areas will use to drain the tank and remove any scale that may have settled on the bottom and make it less efficient. The drain may be plastic and will have a thread that a garden hose will attach to. Make sure you keep a garden hose, or a short hose for this purpose.

If you attach the hose and turn the drain, after the water has cooled, you may notice very little will flow out. This is because the tank has a vacuum in it and needs air to enter the tank for any water to leave it. Your hot water tank will have a "pressure relief valve" located higher up on the tank, which you can open, as long as the water has cooled off. This will allow air into the tank so water can drain out the bottom. If you attempt this with a hot water tank that is operating at its normal temperature you can easily scald yourself, quite badly. So DO NOT attempt this until you are sure there has been no electricity or gas going to the tank and that the water inside has cooled off.

Drinking Water - Rural

People in the country are usually more independent when it comes to water than people in the city. They have wells that supply them with water. Hopefully you're far enough away from industrial pollution and agricultural runoff that your well provides good drinking water. If it doesn't you should look at a system to purify your water, like a reverse-osmosis unit or a water distiller. Water distilling is basically boiling water and then capturing and condensing the steam to avoid any pollutants, so it uses a lot of electricity. (It's not something you could use conveniently off the grid.)

Whether you live in the country or city a reverse-osmosis system will give you very good-quality drinking water. If your city is struggling with water infrastructure-related challenges it may not be a bad idea to have a system like this. You also may want to make sure you have a jug of bleach around and an eyedropper. If you can't boil water in the event of a power outage you can disinfect water with household bleach. Bleach will kill some of the disease-causing organisms but may not kill them all. But it's better than nothing. If the water is cloudy, filter it through some clean cloths or allow it to settle. Draw off some of the clear water and add 1/8 teaspoon or 8 drops of regular unscented liquid bleach for each gallon of water. Stir well and let it stand for 30 minutes before drinking it.

Your well will be either dug or drilled. If it's "dug" it means it was dug out and a concrete wall built around it. Many rural homes have a company come in and drill a well with a large drilling truck. A perforated-steel well casing is inserted which keeps material from falling back into the well while letting water trickle back in to fill it up. Inside the house there's a pressure tank, which pushes the water out of the taps. After water has run for a while in the house the pressure in the tank will drop, and at a certain point the pump will click on and pump water up from the well and into the tank to pressurize it to the required set point. A pump requires a fair amount of electricity. If the power goes out obviously the pump won't work. You may have a few toilet flushings' worth of pressure left in the tank but it will run out. If you have a generator to deal with power outages make sure it will handle your pump. Some smaller generators can't supply enough power when the pump comes on, and you'll hear the generator sputtering if it doesn't quit completely. So size your generator accordingly and test it before there's a power outage.

If you have your own rural water system your main concern during a power disruption is keeping the water flowing. A generator is one possibility and the other is a renewable energy system with battery backup. Again you'll want to make sure your inverter is sized to handle the pump. A pump will come on with a surge and at this point it requires a huge amount of energy. Once the water gets flowing the pump will use less and less energy over time. A good inverter will have a surge capability much larger than its rated capacity to handle this. An inexpensive inverter will not, and you'll know it because it will shut itself off when exposed to a load like this that's too large for it to handle.

Some older rural homes have cisterns, which are large concrete tanks that hold the water which comes from eavestroughs that divert rainwater. In drought-prone areas these are very handy, but you should monitor the water quality if you are going to be drinking it. I will always remember the story that writer Timothy Findley tells in his book *From Stone Orchard* of how badly the water smelled and tasted when they moved into their farmhouse. Further investigation discovered the decomposing body of a raccoon that had died in their cistern. It is always a good idea to have any rural water supply tested by the local health authority.

Water Sanitation

Toilets are pretty amazing things. So are sewers. We can simply flush away the wastes that for much of human history contributed to disease and sickness. We tend to take toilets for granted—until they stop working. Then it's an emergency. When 50 million people in the northern U.S. and Canada were plunged into darkness in the blackout of August 2003, some learned very quickly about toilets. Most cities have backup generators to maintain water pressure, which keeps toilets flushing. Apartment buildings all need pumps because the water pressure from the city will only get the water up to the 6th floor. So people living on upper floors in apartment buildings with no backup generators or generators that didn't work had no water. I remember the television image of people walking down 20 flights of stairs to get a bucket of water to flush the toilet. Sure it was great exercise for a few hours, but if it had continued for days people would have taken to the streets. And it looked as if most people hadn't practiced this regularly enough to be in shape to handle it.

In a city you are pretty much dependent on someone else to provide your water for you. Municipalities have done an excellent job at supplying water reliably for decades, but it's important that you realize this can always change. Anything you can do to anticipate water disruptions is going to make your life a lot easier. If the news calls for inclement weather, have a bath the night before and leave the water in the bathtub. Should you lose water pressure when the storm hits, you'll be able to flush your toilet with that water. Simply scoop out a bucket from the tub and pour it into the bowl and it will take what's in there with it. If it comes to this make sure you don't flush for each pee either. If it's yellow let it mellow. Don't flush it until you've used it two or three times.

The same can be said for your kitchen sink. Keep it filled with water. Have a couple of buckets around that you keep filled with water. If you see large plastic water jugs in the neighbor's recycling box, grab those and keep them filled up with water for emergencies. You can always use it for flushing toilets and washing hands.

Composting toilets are becoming more common in cottage and rural applications but haven't hit the mainstream in urban areas yet, and I'm not sure they will. There are two main types, some which use electricity and some that don't. Obviously for an off-grid application your preference should be the one that doesn't use electricity. Some models also use no water while others use some. If you're considering one, evaluate your priorities for the system and then get a handle on what your usage will be. You may only use it for four weeks in the summer, but if you have eight people there while you're using it you'll need to make sure the toilet can handle the demand.

From a personal hygiene perspective handwashing is probably going to be as good as it gets if you have prolonged disruption to your water supply. You can see where my bias towards rural and renewable-energy-powered living comes in. At my house my water is always available, clean and ready to drink, and it's pretty much limitless. As long as we have sun and wind we have water flowing. In 15 years of living here we've never had a day without water.

People who camp are familiar with the issue of storing drinking water. If you have the large camping jugs make sure you keep them filled up. There are collapsible plastic containers with handles that you can keep at the ready. Aquatank (www.aquaflex.net/) makes lightweight, portable plastic containers that can hold up to 150 gallons. If you anticipate water disruptions in your area, investing in some of these units would be a good idea.

The key today is to prepare for a water disruption and hope it never happens. I'm not advocating having a 45-gallon drum of water in the living room of your 15th floor apartment. I am suggesting, though, that you realize there is a tremendous amount of energy involved in getting the water out of your tap and that in an emergency energy can often be in short supply. Drinking water in cities is maintained with energy and chemicals to make sure it's safe to drink, and during a prolonged power outage some systems may fail. You just need to be prepared for this possibility and not be lined up with everyone else trying to buy bottled water.

Water for Gardening

Rain barrels are essential at your house, and one isn't enough. You need them on every downspout, and on the downspouts where the largest amount of water from your roof is channeled you need a few. Most of the time these are going to be for watering your garden. With the challenges of water shortages in the summer these will be crucial. In an urban environment you should keep them filled as much as you can for as long as you can. Rain barrels will give you days' worth of toilet flushing and basic water uses like dish washing should there be a disruption to your water supply. Be careful during the fall and spring, though, as freezing can crack and destroy a plastic rain barrel.

Water sitting in rain barrels will start to grow algae and get ugly after a week or two of hot weather, which is why you should cycle the water through them. Within a few days of a rainstorm use the water for the garden. Then if no rain is forecast refill the empty rain barrels from your municipal water supply. If you have two 50-gallon barrels this gives you 100 gallons of water on reserve. If they are not being filled with rainwater, after a few days drain one into the garden, then drain the other the next day, and keep cycling them that way. In terms of gardening water this is actually better for the plants anyway because you've taken that very cold municipal or well water and allowed it to warm to the outside temperature, which will avoid shocking your plants. Allowing the water to sit will also help to dissipate some of the chlorine that is in much municipal water.

If your garden is large you'll need to get creative with water storage, especially if you live in a drought-prone area. There are lots of large rigid plastic containers you can purchase. Trying to find a source for used containers may save you some money if you can verify that they haven't had some nasty chemical in them. What about that old pool the kids don't use anymore? Or the larger pool your neighbors had a few years ago and got sick of? Maybe they'd trade it for a couple of baskets of tomatoes when your crop comes in if you can use it as a rain reservoir. That old canoe by the back fence is great for storing water when you turn it upright. I'm always trying to find new ways to store water. The reality is that when you get a summer rain sometimes you'll get way more than your rain barrels can hold, so I'm forever running the extra into our old canoe and kiddy pool to increase my reserves.

The health authorities would climb the walls at my house because of all the water I have sitting around. Reservoirs of water are breeding grounds for mosquitoes, which today can spread West Nile Virus. You should take action to prevent mosquitoes from breeding in your rain barrels. Commercial barrels are covered or screened. Several drops of vegetable oil is said to prevent the larvae from hatching. My suggestion is to keep cycling and replacing the water. Sometimes at my house I'm not as vigilant as I should be with my 12 rain barrels and yes, I find mosquito larvae. But my house is surrounded by ponds and the mosquitoes can breed just about anywhere around here, so it's a losing battle. Luckily the mosquitoes here don't seem to have the virus yet, but you should be careful.

The one thing I make sure of is that my rain barrels are elevated to make it convenient to get a garden hose on them. This helps with moving the water to the garden if it's a distance from the house. It also allows me to conveniently fill up the old kiddy pool and boat that I use to store the excess during downpours in the summer. I live in a very dry area and have sandy soil which dries out quickly. I also have a very large garden, so when I do get a big rain during the growing season I store as much of the water as I can.

My rain barrels are a dark plastic and the water in them will keep longer than it will in things like the kiddy pool. The water turns green pretty fast in there, so I use that water first. I don't wait too long because I find that if I leave the soil too long after a rain the surface of the soil gets hard and will not accept water as readily. So a day or two after the rain I start using up the water that will go bad first, giving the garden a deep, thorough watering.

Water Conservation

It goes without saying that you have to use water efficiently and not waste it. You only have to look at the "bathtub ring" around Lake Mead to see how much water levels are dropping. Most people know not to leave the tap running while they clean their teeth or shave. Many municipalities are mandating low-flush toilets because so much water is wasted with old designs. Up to 40% of a household's water is used just flushing the toilet. The ultimate way to avoid this is with a composting toilet. Since many people, especially urban dwellers, probably aren't anxious to embrace that level of environmental stewardship, a better strategy is to install a low-flush toilet, which uses 1.6 gallons (6 liters) per flush. An even better solution is a dual-flush toilet, which uses just 1 gallon (4 liters) for the easy stuff and 1.6 gallons for work requiring more water. In areas where water is really precious a toilet with a hand-washing basin on the back has been designed, so that the water you use to wash your hands goes into the toilet tank to be used for the next flush. It's a brilliant concept and one that is starting to recognize the importance of this precious resource. This respect for water is something that really hit home to me when we moved to the country, where we have a well and a finite amount of solar and wind energy to pump that water into our home. I keep a plastic jug beside the bathroom sink. When I'm running the water to wash my hands and I'm waiting for it to get hot, I run that cold water into the jug. During the summer I use that water in the garden and during the winter I use it on the woodstove to humidify the air.

We wash dishes in a dishpan so that we're not filling up the whole sink. I run just enough water in the bottom to cover a row of dishes and mugs. Then as I rinse the glasses after washing them, the water gradually fills up the dishpan. By the time I've finished I've used a fraction of the water of a dishwasher and none of the electricity it would use.

New products are coming on the market that also let you become much more efficient with water. With a graywater recycling system by companies like BRAC (www.bracsystems.com) you take the water that's left over from baths, showers, and laundry, filter it, and then use it to flush your toilet. It's such a brilliant concept because that's perfectly good water for toilet flushing and you've saved all the energy and expense of having your city clean and pump you fresh water for the same task.

Places that are severely water challenged, like parts of Australia, have helped develop rainwater catchment systems (www.rainharvesting.com.au). You can drink rainwater but you have to be careful. These systems start with a good gutter screen to strain out leaves and larger items from the eavestrough. Then the rainwater goes through a second debris screen before it goes into a first flush diversion. This brilliant invention lets the first rainwater that hits your roof be diverted until the rain has given the roof a good cleaning. There may be deposits resulting from air pollution and pollen in that initial water, so you divert it and use it to water the garden. When the diverted water reaches the top of the reservoir, the rainwater is then diverted into the main reservoir, which is a metal or plastic tank. The water is then filtered for drinking. When you see systems as well developed as these appearing on the market you know we have both water challenges and solutions.

Treating water like liquid gold is a responsible thing to do for the planet. Anticipating and preparing for a disruption to the water that's supplied to your home is the responsible thing to do for your family. The two goals are perfectly complementary. Once again, what's best for you is best for the planet as well.

18 Transportation

We love our cars and we love to drive. We were all "Born to Be Wild," to jump on our motorcycles and into our cars (and RVs) and hit that open road. It's pervasive in our culture from books, to music, to "road movies." Jack Kerouac's book *On the Road* is said to have defined a generation that was mobile and always on the move. Our whole culture is based on a one-time bestowment of fossil fuels that are now running out.

Since oil peaked at $147/barrel in 2008 we seem to have hit a plateau of the world producing about 85 million barrels of oil a day. Early in 2013 the price of Brent Crude Oil is $110 when there are many indications the world economy is slipping back into recession. Reduced economic activity generally reduces demand for oil, which drives down the price, yet it remains stubbornly high. This plateau we're on could be the indication that we simply can't extract any more oil out of the ground. We've hit peak.

This is all happening as many of China's growing middle class enter the car-ownership market and Tata Motors in India has started shipping its "Nano," a $2,500 car designed to get people off bikes and scooters and into the driver's seat.

The reality in North America is that we will have to rejuvenate our railroads since many of us will begin taking the train more often. Trains are extremely efficient. Unlike a car or bus where you have rubber hitting asphalt resulting in greater friction and therefore requiring more fuel to move the vehicle, trains have steel wheels that roll on steel rails with very little friction or resistance. Compared to the energy required to get a jet off the ground and keep it moving, trains are simply light years more efficient. Rail travel will have the obvious advantage of reducing how much CO_2 we put into the atmosphere. Jets are particularly bad because they release CO_2 high in the atmosphere where it does much more damage than at lower altitudes.

Ultimately the peaking of oil supplies could mean we're going to end up living in villages again. Even in cities people will drive less and trade much closer to home. Our economy is going to return to being more local as oil becomes prohibitively expensive and increasingly hard to come by. For those who were against the march towards an integrated world economy, it appears that peak oil is going to do what a concerted effort on the part of individuals across the planet couldn't, namely reverse the tide of globalization. The message from an individual point of view is that you are going to be traveling less and less, which impacts your decision about where you live. The advantage of a city of course is that with a high concentration of people in one place the economics of transit is much better. Peak oil is basically going to bring about the end of suburbia. Urban design which requires cars as the basis for all transportation is going to become increasingly unsustainable. Having to use your car every day to go to work, attend social and sporting events and go shopping will become prohibitively expensive as the effects of peak oil start to show up at the gas pump.

Europeans have already reacted to this reality by building more compact cities. The smaller distances between major European cities also makes high-speed rail much more viable. For some trips it is now faster to take a train than a plane between major cities. North America has large distances to cover, but we'll have to either do it by train or stay home. We're going to have to take to the extreme the 1980s concept of cocooning—staying home with popcorn and a rented movie rather than going out on the town. We're all going to have to become homebodies. If you like the nightlife this will be a bad thing. If you like the comforts of home this will be a good thing. Regardless of which camp you're in, it's going to require a huge adjustment for a population that has lived with endless and cheap travel to deal with restrictions on its movement. This will be difficult for many people to deal with. Cheap and abundant energy has shaped generations of North Americans. It has allowed families to live at opposite ends of the country or in different countries altogether. I can't tell you how many people I've met lately with older children who live all over the planet. If families continue to live this way they are going to see each other less and less. Trips home at Christmas are going to become extremely expensive. They may get together every other Christmas at first, but as the rate of oil depletion increases they may not be able to afford to do it at all.

I've had participants at my workshops put questions to me like this: "My fiancé lives in Australia. He's close to his family and I'm close to my family in Toronto and we wonder which country to live in." Decades ago this rarely happened because travel was slow and expensive and the majority of the population couldn't afford it. Now with airfare so cheap and so many people going to school and traveling to other countries to explore and work it's inevitable that people from different locations fall in love. The problem is that these relationships are premised on the cheap and abundant energy model, and that model is about to come off the rails. There's not much I can tell them. Whichever place you choose to live you are going to risk one of you not getting to see your family very often. That will be a heartbreaking thing for everyone involved, but I believe it is the new reality. It brings me no joy to inform people of the inevitability of the end of oil, but I believe we all need to be realistic.

We are all going to be staying closer to home. Services like Skype (www.skype.com) that allow us to video chat with people over the Internet are mind-boggling in the amazing technology they provide, and they are going to make it a lot easier for people in these situations. Grandparents in a different part of the country are going to get to watch the grandkids open their gifts on Christmas morning over the Internet. Not quite the same as being there, but better than not seeing them at all. It will also help to eliminate some of those inevitable family squabbles when Uncle Billy has too much eggnog and starts dragging skeletons out of the closet.

The transportation issue is a good reason to consider living in an urban area. Hopefully governments will keep transit working to allow you to move about the city for work and shopping. Larger cities will also be hubs for rail transportation between cities. Some smaller cities may have more difficulty in this department because over time, as truck-based transportation has usurped railways; many rail lines have been abandoned and left to decay. In the rural area where I live all the villages and towns were at one time served by regular rail service. As cheap oil allowed people to own cars the rail lines were abandoned. The rail companies didn't like the liability of people getting injured on their property so many turned them over to governments and organizations for use as trails. While some are for cycling and walking many are for gas-powered activities like ATVs and snowmobiles. The unfortunate part is that as these trails are abandoned by vehicles that require gas which people can no longer afford for recreational activities it will be very difficult and extremely expensive to ever return them to rail service. The railway ties and steel rails are long gone. It requires a huge commitment on our part to resurrect them.

So transportation is going to be much more of an issue for rural dwellers. Or it's going to be an issue for someone choosing to live in a rural community and believing that they will be able to continue to live as we have for the last 50 or 75 years. It is unlikely that you'll be able to live in a small village 45 minutes from a larger city and commute to work each day. Unless you have an extremely high-paying job, the economics of long commutes is going to be increasingly prohibitive for the average North American worker. If you work in the technology field or have a job that requires just Internet access and a computer then telecommuting may work, allowing you to work from home. Perhaps you could spend most of the workweek at home and commute to the office for face-to-face meetings once a week. Many rural communities now have high-speed Internet so there's no reason you can't have face-to-face meetings with co-workers in cyberspace. Our family's Apple laptops all come with a camera installed and we can video chat with our daughters in the city two hours away whenever we want. Saves time, gas, and carbon emissions.

For a while people in outlying communities will have to form carpool systems that allow a number of people to drive in one car to work in the city. There will be a transition period as higher oil prices inspire a more aggressive search for the remaining oil, but we've found all the giant or elephant fields and the new discoveries are always smaller and much harder to get at. During this period it will appear that with some discomfort we'll be able to maintain our happy motoring lifestyles and in the short term it may not be so bad sharing a ride to work and getting to hear about what idiots other people's bosses are too. But in the long term your goal must be to be independent and not require regular driving, anywhere. If you can scrounge up some gas, great, but plan in the future on "staying put." This gets you back to your "where to live" strategy. If it's rural, there is an advantage to being able to walk to town to the grocery store and other amenities.

One of the questions most of us have when confronted with the concept of "peak oil" is, "What's the big deal?" On the surface this is a fair question. Most of us don't understand the ability of fossil fuel to displace manual labor, and that is the big deal. Three spoonfuls of crude oil contain the same amount of energy as eight hours of human manual labor. I know this from my garden. I know how much manual hoeing and shoveling just a bit of gas in my rototiller displaces.

Every time you fill up your gas tank, you're using energy equivalent to one person's manual labor for two years. A barrel of oil equals 8.6 years of human labor. A human lifespan could produce about three barrels of oil-equivalent energy. That is discouraging! Your whole life's work distilled down to three barrels of crude oil. The mind boggles. Yet we've all come to take this little miracle for granted.

I've read that the average suburbanite with an SUV in the driveway has more power at her disposal than the Pharaohs had with 1,000 slaves to build pyramids. The problem is that we just don't appreciate the potential in that magical gas we put in our tanks. The other problem is that we do not really have an alternative. There is no substitute for gasoline. There is no economical, readily available source of energy to replace the potential energy in a gallon of gas. Not the sun, the wind, not hydrogen, nothing.

We all have high hopes for the hydrogen fuel cells that we keep hearing about, but hydrogen is not an energy source—it's an energy carrier. You have to make it from another fuel, right now usually natural gas. So it doesn't offer a solution to our rapidly declining fuel reserves. What about electric vehicles powered by solar and wind? As wonderful as this sounds there simply isn't enough sun and wind power to move around the country the way we're used to. We still need to rely on the existing electrical transmission system which uses coal and nuclear power. The reality is that the grid is pretty much at capacity now so it would be very difficult for a large number of cars to switch to electric.

There are a billion cars in the world and over 200 million cars in the U.S., so the move to electric vehicles is going to take a long time. People keep their cars on the road longer these days, especially during an economic downturn, so you won't find a huge percentage of vehicles being replaced today anyway. Even if automakers come out with electric cars and plug-in hybrids, they will not make up a significant percentage of the fleet for a long time.

In the meantime people are going to continue to buy and burn gasoline and this is going to lead to the decline of world oil reserves, with gas prices going up significantly as supply falls.

We all know that driving is a bad thing for the environment, so using less gas is actually a good thing. Right now if you're considering a new car make fuel efficiency your number one priority. After oil prices dropped from $147/barrel we all quickly got used to cheaper gas prices again. Don't be fooled. This will not last. Remember that gas is going to become increasingly expensive and in some cases in short supply. So you want to make sure your car can travel as far as possible on a full tank. You'll notice I've been saying new "car" rather than "vehicle." Your days of SUV and mini-van driving are over. You need a car. A small, fuel-efficient car. If the kids are a little cramped in the back, too bad. If they don't each have their own beverage holder and DVD player they're going to have to deal with it. Get 'em in therapy. They're going to have to start realizing that they should be grateful to be moving around in this magical, independently powered steel box, which is a miracle unto itself. The days of frills are over. If you're concerned about driving seven kids to a soccer tournament, get to know the local rental car agency. Rent a truck or mini-van when you need one. Don't be driving around in it 30 days a month for the 1 day a month you really need the capacity. Or let the other parents drive those days.

You should also give some thought to a diesel, especially if you have a long commute right now and you don't see reducing it in the near future. Diesel engines are on average 30% more energy efficient than an equivalent gasoline engine, and with newer, cleaner turbodiesel and common rail diesel injection motors, emissions are much improved from the old days of smoke and noise. The problem with gas or diesel fuel in the future is going to be supply. I believe this may be more acute with diesel fuel because you will be competing with uses that may be deemed more essential than your commute to the mall. Trucks that move goods around the country and tractors which farmers use to grow food use diesel fuel. If we get to the stage where there is fuel rationing, it may be harder to get diesel fuel. In the fuel crisis of the 1970s there were stories of sugar being put in the fuel tanks of diesel cars because truckers didn't like having to compete for limited diesel fuel. Luckily cars today have lockable fuel caps.

The upside of diesel fuel motors is that you can run biodiesel in them, which you can make from waste vegetable oil from restaurant deep fryers. While this sounds easy and cheap, there are a number of steps involved to make good-quality fuel. Supply of waste vegetable oil could become an issue as well. As the price of crude oil skyrocketed in the spring of 2008 restaurants that were once happy to have home brewers haul away their waste vegetable oil found industrial waste-oil haulers were prepared to pay for the privilege and in some cases bidding wars broke out to purchase the waste oil. If you're counting on waste oil you could be shut out of the game if crude oil prices rise high enough. An excellent source on biodiesel is William Kemp's *Biodiesel Basics and Beyond*. It's the most complete guide available today on how to make ASTM-quality biodiesel.

You should also consider a hybrid vehicle, which most manufacturers have available. An even better option is a plug-in hybrid. Today's hybrid cars have an electric motor to assist the gas motor, but most of the energy in the car still comes from the gas motor. A plug-in hybrid boosts the battery capacity, which allows the car to run as an electric for a bigger percentage of the time. Some people may be able to run errands and take short trips on electric power alone. The batteries are either charged by the gas motor when it's running or by the grid when you plug it in.

Some people look to motorcycles and see a motorized two-wheeled machine and assume it has fuel economy similar to that of a bike. This often isn't the case. The average fuel economy for a mid-sized motorcycle is about 60 miles per gallon (4.7l/100 km). This is worse economy than most hybrid cars and is comparable to regular gas-powered compact cars, which can carry five adults and provide heating, luggage storage, crash protection, and shelter from the elements. So if fuel economy is your priority don't be looking at a motorcycle.

Scooters are becoming popular in cities but they have the same drawback as a motorcycle in that they require gas. I recently saw a Vespa gas-powered scooter advertised that had only marginally better fuel economy than my Honda Civic, although I'm sure there are models with better gas mileage if you do enough research. The Vespa looks like a blast, but I believe driving it in a city is much more dangerous than driving a four-wheeled vehicle, and without significantly better fuel economy I'm not sure I see a huge advantage to it. If you're just using it for shorter trips in town a Nissan Leaf electric car will give you four wheels on the ground and the ability to charge it with your solar panels or from the grid at off-hours when electricity is cheaper.

The overarching theme is that the days of personal car-based transportation are drawing to a close. How much money you choose to invest in a car should be tempered by the realization that ultimately you may not be able to afford or even locate liquid hydrocarbon (gas or diesel) fuel for it. If you get an electric car you need to evaluate the range of the battery system and remember that if large numbers of drivers switch to electric cars electric utilities will have to raise their rates to deal with the increased demand. While you'll be able to purchase solar panels to charge your electric car you'll have to ask whether this is the best use of this very expensive, high-quality power that you've generated. Would it not be better used powering your home rather than your car? I think you'll find that your home will take priority and that you'll have to find alternatives for transportation.

I provide the following illustration to give you an idea of how much energy our current lifestyles consume. Energy can be broken down into calories, giving us a standardized unit to measure diverse things. A gallon of gasoline for example (4 liters) contains 31,000 calories. This chart shows how much energy is required to drive 1 mile (1.6 km) to the grocery store to buy dessert for dinner tonight. I'm a huge dessert fan!

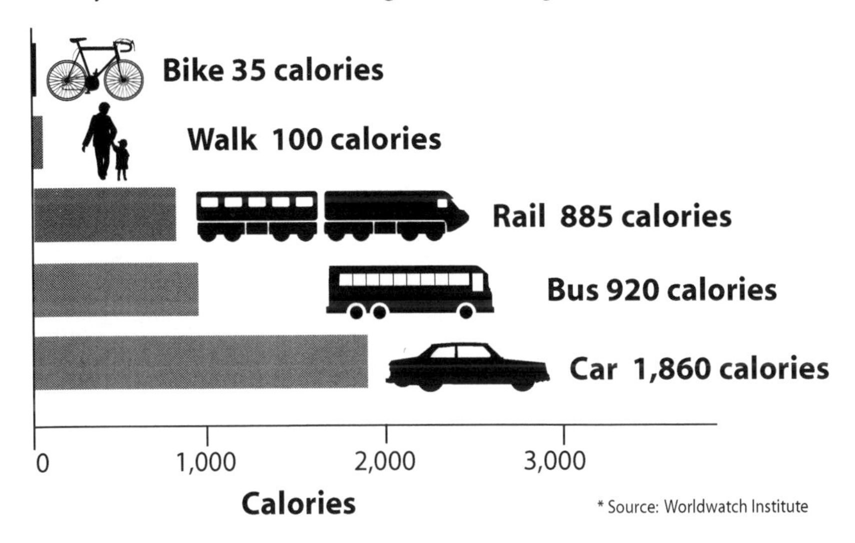

You can see that the most efficient form of transportation available to anyone on the planet is a bicycle! That's good news because it's good for the atmosphere and bikes are pretty affordable for a good percentage of the planet. A bike is almost twice as efficient as walking because it's using those amazing round tires and gearing to accomplish more work with less effort. Those tires make it much easier to move your body weight than just walking. In the case of riding or walking the calories come from food you eat that is converted to energy for your muscles. To ride your bike to the store you need to eat about an apple and to walk to the store you need a banana. Sounds like you're going to be making a fruit salad for dessert!

Then we start using fossil fuels and the efficiency of moving you goes way down. Your share of the energy required to move that fully loaded bus or train is 9 times more than if you'd walked and 25 times more than if you'd ridden your bike. Things get really out of hand when you decide to drive your car to the store. You can see your car uses 50 times more energy than if you'd ridden your bike.

I have not included the calculation for a jet in the chart and it hardly requires mentioning that a jet uses an obscene amount of energy. While most people don't take a jet to the store, air travel continues to grow every year. If you've flown in a jet and heard the engines wind up as you take off you get a sense of the energy that is required to get that 200-ton fully loaded Boeing 767 off the ground. The mind boggles at modern flight. And once you get that sucker airborne, you've got to keep putting the pedal to the metal to get up to its cruising altitude. Then you've got to keep those 200 tons of metal moving through the air, defying gravity. It is difficult to get an average number of calories for air travel because it depends on the type of jet, the distance traveled, the number of passengers on the jet and a number of other variables. The impact of a jet engine burning kerosene or jet fuel and spewing that $C0_2$ and water vapor high into the atmosphere is much worse than the effect of $C0_2$ created on the ground. The atmosphere is much thinner and less resilient to such an intrusion. As George Monbiot says in his book *Heat: How to Keep the World from Burning*: "If you fly you destroy other people's lives." He of course is talking about those who live in low-lying areas of the world that are particularly vulnerable to the ravages of climate change, like rising sea levels in the Maldives and Bangladesh.

I hope Monbiot's statement will discourage you from flying and I hope even more that it will inspire you to get out on a bicycle right away, because bikes are very efficient. Insanely efficient! So you're going to get one. My attitude has always been to buy a cheap bike because bicycles tend to get stolen. An inexpensive bike won't have the smoothest gears or the nicest brakes, but I've always felt that if I have to work a little harder to get the thing moving it's a better workout for me, and I'm way ahead in terms of the energy required to walk the same distance. The three main bike types are mountain bikes, racing or road bikes, and hybrids, which are a cross between the two. Mountain bikes have the big knobby tires and are good if you're going to be riding over very rough terrain like trails or streets that are in really rough shape. City roads can be pretty broken up and with sewer grates and streetcar and train tracks a wide tire is considered best.

Road bikes have very thin tires and are good if you plan on riding a really long distance. The thinner tires mean less resistance and therefore less effort to travel a given distance. The downside to those thin tires is that you are much more likely to take a spill when you hit a pothole or broken pavement. Some are thin enough that they will actually slip between the grates of a sewer cover, although most cities have been changing these over to diagonal ones that are safer for bikes. So you'll be able to ride farther on a racing or road bike but you'll have to be more careful about obstacles on the road.

A hybrid bike, as you can imagine, is a cross between the two. It will have intermediate tires that are thinner than mountain-bike tires but wider than the tires on a road bike. They are better for a long commute in a city with marginal or questionable roads. Many of the newer designs also follow the much older design in terms of having a bigger seat and higher handle bars so you don't have to ride crouched over, which can be hard on your back over a long distance. I like my road bike for long rides and don't mind being stooped forward because it means my body is more aerodynamic and I don't have the same drag and wind resistance. Michelle loves her hybrid which has a seat built for a woman's body and higher and more comfortable handlebars.

I'm happy with a $150 bike and if it gets stolen it's not the end of the world. Our latest purchase was an electric bike and it is the coolest thing ever! Electric bikes have been refined over a number of years and their technology is getting much better. The original electric bikes used lead-acid batteries that were very heavy and had limited range. Our new Schwinn bike, which cost $1,000, uses lithium polymer batteries. The range is 16 miles (25 km) and the battery should take more charges and last much longer than a lead acid. From what I read it's best to store a lithium battery with about a 50% charge in the cold. You fridge would be a good consistent temperature but there's never room in ours. We have had this bike now for 5 years, each year storing the battery on the bike outside in the barn and each year the battery bounces right back and we've had little deterioration in its performance.

This bike has three modes: you can just ride it yourself, have it assist you as you peddle, or have the DC motor take over and do all the work, like a motorcycle or moped. These bikes are great because they let you get some exercise but they also help extend your range with the motor. If you like to pedal, an electric bike helps you maintain a much higher speed so that you're more likely to ride than drive. I find it amazing going up hills because you continue to pedal and even though you lose a little speed you can keep up a good head of steam on a fairly steep hill. It's completely brilliant!

If you were to use an electric bike to commute to work and your workplace has no shower facilities you might want to use the motor on the ride in so you don't get too sweaty and then use mostly pedal power on the way home to relieve the stress of the day. When you get home from a ride you can charge the battery by plugging it into a regular outlet. Since we live off the grid the electricity comes mostly from our solar panels, so we have a solar-powered bike! How cool is that? In fact this would be a perfect way to get you into solar power. Put up a panel and small inverter and you can use it to charge your bike. This is where solar-powered transportation becomes a reality. Energy consumption in vehicles in large part comes down to weight. A bike is very light, so using solar power to help with transportation becomes a reality if we're talking about something like electric bikes.

If I were living on the electricity grid, the electricity to get me into town and back would cost me about a dime. Not bad. Sure it's not optimal on a snowy day, but for a good chunk of the year an electric bike is an excellent alternative to a car.

My neighbor who rides a motorcycle asked about the noise of an electric bike. I assured him an electric motor is virtually silent. Then I realized he expected it to roar like a Harley. The best I could offer was to attach a baseball card to the front fender with a clothes pin and letting it slap against the spokes as you ride. It won't attract biker chicks, but it will attract enviro chicks. Was this an appropriate thing for a feminist to say? I think not.

There are many electric scooters coming onto the market and again you could use solar power to charge these. One thing to remember with some scooters is that in certain jurisdictions they are considered a vehicle, like a car, and therefore you need to follow standard licensing and safety regulations. Some scooter makers try and get around this by including pedals that allow you to pedal in a worst-case scenario, but they are heavy and the pedals are an afterthought, often very spread apart and uncomfortable so that in reality you couldn't ride it very far.

One final possibility for transportation is horses. I can hear you groan and say, "We're not going back to *Little House on the Prairie* times I hope." Well, no one is going to force anyone to wear crinolines and bonnets while feeding the pigs, but if you live in the country riding a horse can make a lot of sense. In the rural community where I live you occasionally see someone who has ridden a horse into town for the novelty of it. There are lots of horses throughout the country right now and many of them spend sedentary lives loitering in horse barns and paddocks, waiting for their owners to ride them on the weekend, when it's warm, and there are no bugs, and there's no chance of rain, and there are no sales on in town.... so we might as well put them to use. They love being needed and used and if you're far enough out why not ride the horse in? As the price of fuel rises I have no doubt that the most creative retailers in small towns will set up horse "parking spots" with water and hay and those hitching rails you see in all the old westerns. It's a chicken and egg thing right now. Most towns don't have places to comfortably tie up your horse while you're in town. So if you build it, they just might come.

Now the purest environmentalists out there will argue that a bike is still a better way to get yourself into town because those wheels and gears are so much more efficient than four legs carrying a 1,500 pound horse. This may be true, but as a gardener who doesn't ride horses I can tell you I love horses. It's the end result of all that hay munching that I love for the garden. In an area like ours with marginal soils, hay is one crop that grows exceptionally well. So turning sunshine into hay and then having the horse turn that hay into manure to put on my garden makes perfect sense to me. I need to replenish my soil, and while I use some green manures there's nothing like composted horse manure.

Your challenge if you decide to use a horse for some of your transportation needs will be feeding it. Even though you don't have to worry about purchasing increasingly expensive liquid hydrocarbons like gas and diesel, you'll still need fuel. Hay will require you to have either the money to purchase it from a farmer who grows it or the acreage to produce it yourself. You would be surprised at just how much land you need to feed a horse. Depending on how much you're working it a horse might require five to ten acres of hay to support it during the year. As the price of food increases farmers are going to turn some hay fields into commercial crops for human consumption, which will raise hay prices. If you have enough land to support horses great, but you may be similarly tempted to turn that land into crops for human consumption. There are so many variables in this calculation that I can't attempt to make a recommendation on whether or not to keep horses. Many people love horses and if you're putting them to work all the better. If your land is poor, improving it with their manure is an excellent idea. For centuries horses displaced huge amounts of human labor, until fossil fuels became available and displaced horses. Looks like we're going full circle and the term "horsepower" will no longer refer to the 400 horses under the hood of an overpowered sports car but to the amount of energy that big, beautiful creature out in your paddock can provide.

One thing that is certain is that the future of transportation won't look like the present. I grew up watching *The Jetsons* and assumed I'd be zipping around in a cool little space car by now. While we humans may one day possess this technology it's a long way off, and in the meantime our predominant form of transportation relies on fossil-fuel-burning internal combustion engines that have a limited future. The earlier you start preparing for this the better. The sooner you decide to make sure you don't require a long commute to work and commerce the better. If you are going to drive, use the absolutely most fuel-efficient car you can find. As a backup plan buy some bikes. Buy a mountain bike for off-road and a road bike for longer trips. Start getting in shape. Start exercising those long-dormant muscles that you don't use when you step on a gas pedal but come in handy for pedaling a bike and moving your mass a distance on the most efficient form of transportation there is. Then once you've got yourself back in shape, reward yourself with an electric bike to increase your range and displace your car miles more and more. Make its purchase part of your "Green Energy Plan" so that once you get your solar electric panels installed charging up your electric bike will be one of their first uses.

The winds of change in our endless happy-motoring lifestyles are blowing hard. They could very quickly turn into a whirling tornado of chaos for people who are dependent on cars for their livelihood. Now that you know the big wind is coming you can go down to your storm cellar and emerge in your spandex, gel-cushioned cycling shorts with your new bike held high over your head like the environmental superhero you are! Better for the coming challenging times, better for your health, and better for the health of the planet!

19 Health Care

Good health is something that's very easily taken for granted. Sometimes you'll read about or see a TV show or movie about someone with a diagnosis of terminal cancer and you'll vow to yourself, "You know, tomorrow I'm going to get up and be grateful that I'm healthy!" And often the next day you do, but then the next day you have that meeting with a stressful coworker on your mind, and the day after that you wake up with a headache because you drank too much coffee yesterday, and before you know it you're back in the rut of "just getting through every day."

Americans are well aware of the cost of health care and the cost of health insurance. Canadians have been shielded from this with universal healthcare but increasing costs end up using a greater and greater portion of government revenues and ultimately in the not too distant future it will become unaffordable in its present form. This all comes at a time of an aging population and fewer and fewer workers with high paying jobs able to support the system. All North Americans can count on a reduced level of healthcare services provided to them in the future.

Much of this book has already laid the groundwork for you to improve your health. That huge garden you're putting in is going to require lots of physical effort to till and weed and water. It's going to require you to expend lots of calories growing your own food. That food you grow is going to be mostly vegetables and they're going to taste better than anything you've ever eaten from a store. Some nights you're going to sit down to a feast of fresh vegetables and they're going to be so tasty that you're not going to slather them in butter and coat them in salt. You simply won't need to because they'll taste out of this world. Coming directly from your garden to your dinner plate, they'll be full of the micronutrients and enzymes that commercial food lacks. What your vegetables will lack is traces of pesticides and insecticides and herbicides and fungicides and all the other chemicals that don't help your body in any way. They help the farmer compete in a food system gone mad, but they don't aid in your good health.

Chances are you'll be eating less animal protein because it will become expensive as all the input costs such as grain increase. We have known for a long time that a diet high in saturated fat is not healthy for us. We also know that many cancers and other health problems like diabetes are caused by poor diet. So moving towards a plant-based diet may potentially reduce how much contact you have with the healthcare system.

So many of us have an image of farmers decades ago eating huge amounts of animal protein to give them the stamina to do the exhausting levels of work required on a farm. Today, for most of us, including farmers, fossil fuels have displaced that human effort. The problem is that many of us still continue to eat as if we were working in the fields all day. If you spend the day doing manual labor and burn 4,000 calories a day then by all means eat whatever you want and as much of it as you want. But most of us don't put in that kind of effort. And while we should be basically limiting our calorie intake to 2,000 many of us eat far more than this. It's calories in calories out. If you eat 2,000 calories a day and burn 2,000 calories a day you're going to maintain your weight. If you eat 3,000 calories a day and burn 1,500, you've got a problem. And for so many of us the extent of our exercise is the walk from the house to the car, from the car to the office, from the office to the restaurant for lunch, and then the same in the reverse.

The concept of restricting calories is pretty basic to most diets, but it's not one that I want to emphasize. One of the keys to helping us get healthy, though, is an awareness that many of us have been conditioned to larger serving sizes. Over the last few decades the volume of food that is served to us at restaurants has steadily increased. Of course this has happened as we have been doing increasingly less calorie burning thanks to our longer commutes and generally busy lives. So you should try and start eating less of everything, and one of the best ways to do this is to replace your regular-sized 10" dinner plates with smaller 7" or 8" dinner plates. This sounds pretty basic, but the visual connection between your mind and body when it comes to food is very complex. Putting a smaller volume of food on a large plate looks to your mind like deprivation. Putting that same volume of food on a smaller plate, which makes the plate look full, has a much better effect on your visual clues about being full and satiated. I know it sounds too easy, but give it a try. Find a second-hand store and buy some intermediate-sized dinner plates. Put a reduced serving on the large plate and see how "wanting" it leaves you feeling; then put exactly the same volume on the smaller plate. You will notice a difference. And then eat that smaller serving slowly and enjoy every bite. When your plate is clean, don't go back for seconds. Or if you must, go back for more beans and cauliflower and skip the higher calorie foods.

I do not need to list the stats on the general health of North Americans. Or the number of overweight or obese people. Not only is there a huge diet and weight-loss industry, there are TV shows focused on losing weight. It's an epidemic. In fact I'm thinking about taking this book and republishing it under the name *Cam's Guide to Weight Loss and Optimal Health*. The key is getting more exercise and moving towards a calorie-limited plant-based diet. I add "calorie-limited" because you can still eat way too much on a plant-based diet. Heck, a meal of pop and chips would count as "plant-based" since the pop is mostly high-fructose corn syrup, and a slab of black forest cake for dessert would qualify too. That's not what I'm talking about. I'm talking about the move to a diet based on starches— rice, pasta, potatoes—with lots of vegetables on the side.

Now that you're aware of the fact that the world has hit peak oil and that we are on the downward side of the curve of world oil production, you're going to seriously consider your mode of transportation. Since you now have the bike you're going to start using it for more trips. Once you install the carrier on the back, you can use the bike for those quick trips to the store for small purchases. You're also going to be walking more. As the price of gas goes up you're going to start walking to the bus stop. The bus takes you to the train and once you get downtown it's a ten-minute walk to the office. Getting your body in shape like this is going to help you weather the wrenching shock of price spikes in oil that will become the norm. Yes, you may still own a car for a while, but if you're not reliant on it then these price shocks aren't as jarring. Maybe this year you'll take the train home for the holidays rather than driving.

Some people who end up in a rural location may begin heating with wood and actually cut it themselves from their property. Country wisdom dictates that "wood warms you twice." First when you cut it, then when you burn it. My experience has been that wood warms me about nine times. First when I cut it in lengths. Then when I drag it with a sled to where I can get to it with the truck in the spring. Then when I load it on the truck. Then when I unload it and put it on the sawhorse to "buck" or cut into woodstove-sized lengths with the solar-powered electric chainsaw.

Then when I split it into stove-sized pieces. Then when I pile the firewood to dry in the summer. Then when I move it into the woodshed at the end of the summer. Then when I carry it into the house in the winter and put it in the wood box. Then when I take it from the wood box to the wood stove. Perhaps I overanalyze, and I am certainly not an "efficiency expert" because there are probably some steps I could cut out.

But I choose not to. I love heating with wood and I love cutting it. I sweat at just about every one of the steps and sweating means I'm getting exercise and burning calories. And that's a good thing for my health. And after a day of firewood cutting I have a slightly bigger piece of black forest cake. Using the calories in calories out equation, I can pig out because I burned it off!

What keeps you healthy is what goes into your body for energy and using your body constantly to do useful work which keeps your mind challenged and makes you more positive and more likely to ignore a little muscle ache and get out in the garden and hoe some weeds.

There is no doubt that as we age we can't accomplish as much and start slowing down. But that doesn't mean we should park ourselves in the garage and throw away the keys. It means we need to work smarter. It means maybe we pay that young kid down the street to do some of the heaviest lifting in the garden and we take on a little more of the "management" work. We all like to joke about the common image of 85-year-olds from Sweden spending the day cross-country skiing, but there is some truth to this. In his book *The Blue Zones: Lessons for Living Longer From the People Who've Lived the Longest*, author Dan Buettner researches the places people live the longest throughout the planet. His first observation is that the people who live the longest eat a plant-based diet. They're not necessarily vegetarian, but meat is an occasional thing rather than the default. This is good news for you if you really would miss that turkey on the holidays. While there are lots of excellent substitutes, you don't have to give it up completely.

Buettner also finds that the people who live the longest often don't have access to all the accoutrements that make our lives so easy. All those machines and appliances are actually making us less healthy. This all comes back to my "Three Benefits Theory." Using a push lawn mower is better for you, better for the environment because you're not burning gas, and, since you don't have to buy that gas (or the high blood pressure medication), better for your personal independence. Lots of the people in places like Costa Rica where there is high longevity don't have cars;

they use bicycles and walk.

The more you start acting as if we've run out of oil, the better shape you'll be in and the less you'll need our health care system. It will also help you mentally deal with the inevitable while you still have the option of using the system when you have to. If you have chosen to live across the country and away from your immediate family and a parent gets sick, it's important to ask yourself what you would do if you couldn't hop on a plane and fly home. This time you have the option. Maybe next time a super spike in oil prices will put a flight home out of your price range. Then you may need to evaluate where you've chosen to live. If family is important to you then migrating closer to home is something you should consider soon, while you can still afford things like a cross-country moving company. And family is important. Studies show that another key to mental health and well-being is a sense of belonging, to a family or community. Sometimes this is a church, sometimes it's a sports team, but a sense of community is crucial to good health. This way you're not in it alone. You have a support network. For some people that takes the form of their family, but it can take many other forms.

If you don't belong to a network or community now some of the recommendations I've made in this book may help you find one. Let's say you live in an apartment in a city, which can sometimes be pretty isolating. Well, you know you need to start a garden, so track one down. Many will be community gardens which are coordinated by volunteers. There will be a board or committee that meets regularly to deal with issues that come up like dealing with the property owner or neighbors and security in terms of making sure that members' food is staying in their plots. So you should join the board. It's amazing how a common cause like this will bring people together. Some of the meetings may be held at local eateries. You might take on the job of surveying members of the garden, which will involve meeting and chatting with them while they work in their garden plots. You'll be surprised at how quickly you'll find people with similar interests.

I'm not suggesting you go out with the goal of forcing yourself on other people in the hope that they'll accept you. I've been trying this for years and ended up in the bush four miles from the nearest human being. But volunteerism has numerous benefits and finding yourself part of a community is one of them. It's good for your health and an excellent way to share information about dealing with the changes that are happening.

One of the benefits of you growing your own food is discovering how much better-tasting food is that is grown without the chemicals that occur in pesticides and other toxic compounds. This may lead you on a journey to discover just how many potentially harmful chemicals you encounter on a daily basis and how many of them you can eliminate. Often there are simple and natural alternatives that are far superior for your health. With others, like plug-in scent dispensers, removing them from your home will save electricity and eliminate chemicals that the products pump into your air. As you start going down the list of how many of these products you don't need you'll find you can expose yourself to a much lighter load of potentially hazardous chemicals. When you discover that the reason potatoes look so nice in the store so many months after they are harvested is that they are sprayed with a disinfectant you would hesitate to use in your house, let alone eat, you'll find that organic potatoes look more appealing. Better yet, you'll discover that the ones you grew that may have a few scabs on their skin and don't look as if they'll win any awards at the local agricultural fair taste just as good if not better than conventionally grown potatoes and will also reduce your exposure to something that would not be considered a health promoter.

So it's time for you to start being your own health care control board and analyzing what you put in and on your body. The information is usually out there and easy to find on the Internet. Try and track down sources that are as impartial as possible, but don't assume that the information you get is perfect. Sometimes what's best to default to is just how close to nature the product is and how many chemicals have had to be modified to accomplish the task. Never assume because the government allows a product to be sold that it's healthy and non-toxic.You need to take a proactive approach to your health to ensure that you minimize your need for the people in the white lab coats.

Comparing this book to others in the field, you'll find that many other books deal with first aid. Some of the more "extreme" think it's important that you know how to deal with a gunshot wound. But I'm not going down that road. You should start building a medical library and good first aid books should be part of it, as should books on herbal remedies and natural healing techniques. If you develop an ailment after losing your job and health insurance, wouldn't it be nice to find that brewing a tea from a plant you can grow in your garden can help relieve the symptoms? Our home is surrounded by poison ivy, which periodically one of us comes into contact with. Another plant that grows well here is jewelweed or touch-me-not, which is said to be a natural way to relieve the discomfort of poison ivy. Michelle has a tendency to get poison ivy in the spring when she's planting the garden, but jewelweed doesn't flower until the summer. So last summer she picked and froze some of the jewelweed so she could use it in the spring!

There is a whole world of natural health out there that doesn't require you to send your hard-earned dollars to pharmaceutical companies. You'll need do some reading and some research, but it will be worth the effort. Illness is sometimes just something being out of balance in your body, and you can often find a way to bring your body back into balance without the trauma of a prescription. When I see prescription drug ads on TV that say, "Side effects may include dry mouth, heart murmurs, night sweats, vomiting and diarrhea, temporary drop in blood pressure...," I start wondering if the problem you're trying to cure doesn't sound way better than the side effects of the drug.

Herbal and homeopathic cures may not work and you may still need to consult a doctor, but if the symptom doesn't seem life-threatening I believe it's worth a try. Or just waiting to see if your own immune system can handle it. Remember, that new diet of fresh fruits and vegetables and your new exercise regime of walking to the store and riding your bike to work have made you much healthier and your immune system much more robust. So give it a day or two and see how your natural defense mechanisms work before you call out the big guns.

In the future we may not have a viable healthcare system and profitable pharmaceutical companies to offer you solutions, so you'll be spending more time trying to find your own.

From Part II you should now have a good first aid kit in your home. It's always a good idea to take a CPR course. While your family will be much less likely to need CPR with your new low-fat diet and exercise regime, you may have visitors who need it. If you live in a rural area it might not be a bad idea to have something like Benadryl around in case someone has a reaction to a bee sting. Benadryl, which you can buy over the counter, is an antihistamine which can help if someone is having a mild reaction to an insect bite or sting. In extreme reactions you may want to have an EpiPen, which is an autoinjector of epinephrine (or adrenaline) for use when someone goes into anaphylactic shock. While you would hope to never have to use this, in a worst-case scenario it would be good to have one around. The pens cost about $100, are good for two years, and should be replaced after that time.

If you live close to a hospital emergency room an EpiPen may not be that necessary, but for people in rural areas they are a good idea. I cannot tell you the number of times I've rolled over old logs or hay bales and disturbed a hornet's nest. I had about four of them descend on my ear once and man did it hurt. Anaphylaxis can occur if you've been stung a number of times in the past. Your body's immune system becomes sensitized to the allergen, for example, bee sting toxin, and even though you've been fine in the past you may suddenly react. Allergies to peanuts and drugs like penicillin can also bring on these events. If the ambulance is a good drive from your place, consider an EpiPen.

While you can't be prepared for every possible health emergency some supplies and books will help. You can always run to the Internet but in an emergency it may be difficult to find the most reliable source of information quickly. There are many excellent websites which can help you deal with some of your health issues. In fact some doctors are finding they are consulting with patients who already have a good idea of what's wrong and what the solution is. If a doctor prescribes a certain drug the response is, "Well wouldn't XYZ be a better option?" We're entering a new age in health care. Many of us are starting to realize that while doctors have spent many years gaining knowledge it often has to be broad-based and they therefore may not have expertise in the problem you're having. So you have to become proactive. This is good conditioning for the days when you can't afford a doctor or the health care network is just not there to serve you so well.

This isn't necessarily a bad thing. Humans have always relied on natural remedies for health problems. They have also relied on healers who focused on less invasive procedures. Childbirth has always been incredibly dangerous for women, but midwives for centuries used techniques of massage and strategies to move babies that didn't want to make their entrance into the world. It's only recently that the North American medical establishment has begun to recognize what a valuable contribution midwives can make to the delivery process. A woman who has worked with a midwife and delivered her baby at home is already on the road to realizing that we don't always need the high priests of the medical profession. They are wonderful to have and in an emergency those of us that can afford medical care are extremely lucky. But the system is starting to strain under the weight of a bloated bureaucracy and a population that's getting older and fatter and less healthy. Your best bet is stay out of the system as much as you can.

Keys to good health:

- Move to a predominantly plant-based diet.
- Exercise every day.
- Become part of a community for support.

Start critically evaluating everything that goes in or on your body. If you can't pronounce it and don't understand how it works, maybe you shouldn't be using it.

Spend some time every day meditating, whether it's yoga or hoeing your garden. Switch off the chaos of the real world and zone out in a natural environment surrounded by green.

20 Safety and Security

I know what you're thinking. This chapter is a no-brainer. Safety and security? Get a gun! Well all right, yes and no, but at least let's look at some options. Lots of people already have guns, but I get a sense that a lot of people reading this book won't own a gun. They haven't had to, and they don't like the idea of us all arming ourselves. And a lot of the people who have already heavily armed themselves are never going to read this book anyway. And clearly, lots of Americans have firearms.

It's estimated that there are 200 million firearms in the United States. More than 50% of Americans say they own guns. That's a lot of guns. And the numbers are climbing quickly. Guns and ammunition sales went up when President Obama first got elected and then again after the Sandy Hook school shooting when the threat of new legislation restricting certain guns might be enacted.

For years I have been giving my Thriving During Challenging Times workshops and I have suggested that there are "soft landing" and "hard landing" scenarios. The soft landing scenario suggests that although we do have many challenges governments, businesses, and individuals will be able to make the changes necessary to deal with these multiple converging problems. The "hard landing" scenario has a more urgent plan of action and it's the one people don't like to think about too much. When I suggest that if governments can't deal with all the challenges, and if businesses don't adapt fast enough, we're all going to be dealing with a radically different reality, my recommended course of action will be much more aggressive. One of the things I discuss is personal security and it's one of the things people want to think about the least.

It's not something we want to think about, but unfortunately if you take all the problems I've discussed in the first part of this book to their logical conclusion, there's a distinct possibility that our lives will in fact change radically, and our neighborhoods won't be as safe as they are now.

Countries all over the world have already experienced low-level instability because of the worldwide economic crisis, so it only makes sense that the military must consider it a possibility here. While none of us likes the idea of the military being called and marshal law being declared, the challenges governments face today are so severe that members of the military and intelligence communities are now discussing it in public. They're concerned, so you should be concerned too. I'm not suggesting that you be concerned to the point of immobility; I'm merely suggesting that regardless of how loathe you are to think of such a possibility, you need to start considering how to protect yourself.

A good place to start is with your neighbors. Groups like Neighborhood Watch are a good starting point to get to know your neighbors and share concerns about security issues in your community. A sense of community may be one of the strongest weapons you have to feel safer wherever you live because you'll know people are looking out for one another. It's one of the advantages of intentional communities and places where multiple families choose to live communally. But if that's too extreme for you, your neighbors will be key to your personal security. It's time to fight your shyness and get to know your neighbors. It's time to have a block party. This year get everyone to bring their fireworks to the middle of the street and have one big display. Or have a street garage sale. There's no better way to meet your neighbors than by buying their old roller blades at a garage sale.

Another good idea to increase your security is to acquire a dog. If you're a little dog person, the unfortunate reality is the bigger the dog the better. Maybe outfitting your Chihuahua with a little voice modulator that makes his bark deeper and louder will help. Well, I'm not sure that exists. Even just having a sign that says "Beware of dog" can help. We live in the country and have a medium-sized dog, but he barks vigorously at strangers' cars. It always surprises me how intimidated people are by our dog because he is a sweet, lovable dog and I doubt he would hurt a fly. But I've seen couriers refuse to get out of their trucks until I called him away, so it seems to work.

Country properties sometimes have long laneways, so a remote sensor that warns you when someone is coming down the driveway is a good idea. It just removes the element of surprise. There's nothing worse than being engrossed in a book or the garden and suddenly having someone standing behind you.

You might consider having a home alarm system installed. These range from inexpensive window warning devices to full-scale remotely monitored systems. Obviously the more functional your alarm, the more expensive it will be. And if you are going to have a security firm tied into your system you have a higher purchase price and the ongoing cost of support. I can't make a recommendation; this will depend on where you live and how secure you feel your neighborhood is. But the reality is that there are desperate people out there and desperate people do things they may not otherwise do. You have to decide if you're comfortable with someone breaking into your home. It's not a good feeling to come home to find that someone has gone through your stuff and taken things you worked hard for. After experiencing a break-in a few years ago we took steps to avoid a repeat in the future.

You may wish to keep more cash on hand and actually have some precious metals in your home in case you can't get to it during a "bank holiday." You're used to having valuables like jewelry around, and now you're about to up the ante. It may be time to have an alarm installed and working and well publicized on your windows to reduce the likelihood of a break-in, regardless of how well you've hidden the silver coins in the old paint cans in the garage. You may even like the idea of an alarm system for when you're in the house in the event of a break-in. Much of this will depend on your income and your comfort level.

Your comfort level will be elevated if you have some self-defense training and now may be a good time to do this. You may not have time or be inclined to put in the effort to become a black belt, and if the intruder you encounter has a gun, it may not make any difference anyway. But some training will give you basic self-defense strategies that will help you be more at ease. One of the key benefits may be the level of confidence it helps you exude in stressful situations. Research shows that criminals will pick victims based on how they walk and what sort of vibe they give off in terms of their self-esteem. If they see someone in a parking lot looking unsure and nervous they are much more likely to target them than the person strutting across that same parking lot as if they owned the place. If you start walking with a "You want a piece of me? Then bring it on" swagger, you are much less likely to end up the victim. Criminals don't want a confrontation. They want the upper hand and if it looks like you're itching for a fight they'll be more likely to wait until the next person strolls by.

These are all excellent strategies to deal with your and your family's security. You must start thinking about them whether you like it or not. And one thing you must consider is a gun. There, I said it. You might not have wanted to hear it, but I've got to discuss it. I grew up shooting a BB gun and a .22-caliber rifle, but when I lived in the city I had no exposure to guns other than to notice them in the holster of the cop standing in line in front of me at the coffee shop.

Then I moved to the country. We feel very safe where we are but we are three miles from our nearest neighbor. We had several incidents with people arriving at our house whom we weren't particularly comfortable with. One evening a car pulled into the driveway and the driver told me that a truck was on fire near our place. His cell phone didn't work so I called the fire department and then went down to investigate. When I got there the truck was pulled off the road and completely engulfed in flames. If there was someone in the cab it was too late and I frankly was not comfortable being there alone. I don't know if you've come upon a truck on fire on a deserted rural road, but there was nothing in my upbringing that made me particularly well suited for this.

Later it was determined that the truck had been torched for insurance, and that was about the time I bought a gun. One of the police officers I spoke to said, "You know you're about half an hour from the detachment in the south, and half an hour from my detachment in the north, and when you call in from here and you're right in the middle it's sort of a toss-up as to which one of us responds." His message was pretty clear to me. Whereas in the city it would take three or four minutes for a cop to show up at my place if I called 911, here it would probably be half an hour, so for half an hour I'd be on my own. If all the officers were on calls elsewhere, it could be longer.

I do not hunt but I also feel an obligation to protect my family, and in my rural location a gun is simply the best option. I took my firearms course and got my license and bought a shotgun. The beauty of a shotgun is twofold. First, it sprays a wide pattern of shot, which means you don't have to be deadly accurate. A friend told me that in the prison where he worked the guards in the towers at one time were armed with rifles. They rarely needed them but when there was a disturbance guards were usually fairly stressed, their hearts racing, and they were not accurate in their shots. So their rifles were replaced with shotguns because as long as they shot in the right direction there was a fairly good chance they'd hit something. When you confront someone and you have a shotgun, they know the drill.

The second benefit of a pump-action shotgun is the sound it makes as you move a shell from the magazine into the chamber. You can load three to five shells into the magazine, waiting to be used. When you're ready to shoot you pull on the wooden slide or fore-end and it takes one shell into the chamber ready to be fired. Doing that makes the very distinctive "che che" used-shell-out, new shell in sound. You know the sound; you've heard it in movies. Arnold is constantly doing it in *The Terminator* movies. It is a sound that intimidates. It says, "The person at the top of the stairs has something that will do me a lot of harm if I choose to go up them." And with a shotgun, after that first shot you can quickly pull the next shell into the chamber ready to fire. The person at the bottom of the stairs basically has to ask himself, "Do I feel lucky today?"

When you make that distinctive cocking sound with the shotgun, it's virtually impossible to tell if a shell has indeed been loaded, so if you have someone in your house who refuses to fire a weapon, they can still cock a shotgun and point it at an intruder and that intruder still has to ask the same question, "Do I want to risk the consequences of them pulling the trigger, or do I want to leave quickly?"

Another prison story involved a fight in the yard. The prisoners were all getting pretty worked up. One of the guards simply cocked his shotgun and all 600 prisoners dropped to the ground.

Even though I don't hunt I am becoming much more comfortable with firearms and I would suggest that if you do purchase one you become very comfortable with it. There is no sense owning a weapon and fumbling with it when you need it. Holding a weapon is a very scary thing for me. I do not like the feeling of owning a machine that could easily take someone else's life. Heck, if I don't handle it properly it can take my life, so I'd better know what I'm doing.

A shotgun can be a very intimidating thing for anyone. I have a 12-gauge shotgun which just about knocks me over when I fire it. I have to brace myself as if I'm standing in a hurricane-force wind to stay balanced. A 20-gauge shotgun would probably be a better option for a woman if you're thinking of a long gun.

Another option for a long gun or rifle is a ".22". A .22 Long Rifle is probably the most popular gun in North America. It is inexpensive and has recoil that is very mild. And .22 ammunition is very cheap. It is a small game rifle so would be suitable for hunting rabbits, groundhogs, raccoons etc. It would not be a good gun for a coyote. You can use the basic iron sight that comes on the gun or you can add a telescopic sight.

If you're new to guns a .22 is an excellent place to start. It's easy to get used to firing and since the ammunition is inexpensive you can take lots of target practice. It's a lightweight, easy to fire rifle that allows you to get comfortable shooting. It would not be considered an optimal gun for personal protection though because it doesn't have the stopping power of higher caliber rifles and shotguns.

Another option is a handgun. These are banned in Canada but can be purchased and owned in the U.S. I will not debate the merits of handguns being legal or not. I will simply say that if you are in a state that allows handguns and you feel your security is at risk, there are some people who feel a handgun offers a solution. Just as I suggested with the purchase of a long gun, join a gun club or shooting range and use the gun regularly. You don't have the luxury of a wide coverage area with a handgun. You fire a bullet and it has to be aimed accurately to have the desired effect. If you have the gun, know how to use it, be confident that you'll be deadly accurate when you do, and hope and pray you never need it.

If you own a gun make sure it is properly stored in a secure location that meets local legal requirements. You may have to store the ammunition separately from the gun. I know that my legal storage requirements mean that I need some warning if I'm going to use it. We recently had a black bear in the backyard. It was wonderful to see and luckily the dog was inside and didn't notice it, which gave Michelle and me time to admire it. If someone were being attacked, though, and I had to use the gun in self-defense, I would hope that the bear was just waking up from hibernation and was slow and dozy, because I'd need a few minutes before I could access and load my gun to scare it off.

Guns are a contentious issue. In a perfect world no one would need them. Today many people feel it's important to own them. In a future with resource challenges and dislocation they could become an essential part of being able to protect the resources you've set aside for a rainy day. We live in a time of abundance. In a time of scarcity we may not enjoy the comforts that the rule of law provides today.

There are lots of other more novelty ideas for security. One is that you should marry into a big family. That means more strong backs for the fields and more security if things get a little out of hand. You could always build an earth-sheltered home and make sure the front faces away from the road so that no one notices you're there. You could buy 5,000 acres in Montana and have the house in the middle of the property with a nice big "buffer zone" all around.

The best security I think comes from living in a "community," a place where people identify with the community and help each other out. I remember a discussion on the *Life After the Oil Crash* website (www.lifeaftertheoilcrash.net) about living in the bush with a gun. The logic breaks down if four people confront you. Many people suggested you would therefore be better off living in a community of 30 people. And of course the response was "Yes, 30 people with guns."

I think we will ultimately be returning to a more local economy where we know the people we trade with very well. This local economy will exist in our community and it will form tighter, safer bonds. While I think it's human nature to want to help out other people less fortunate than ourselves, people will be expected to work and contribute if they are able. Just taking from someone else won't be acceptable in this local economy. I believe we will return to a more local economy where members who trade in the economy will create a much more livable society and a much safer place for all its members.

If in the meantime you haven't found the community, get in shape, get a dog, get an alarm, and/or get a gun.

21 Money

Writing a chapter on financial independence after (or during) the most severe economic crisis of our lifetime takes on great significance. You want to offer some omnipotent secret to financial success, some tip or strategy that will allow readers to profit from the collapse and retire to the beach when it's all said and done. If you're looking for a magic bullet or a miracle drug that is going to sort out this financial mess, you've come to the wrong place. There simply isn't one. If there were one single piece of advice I could give it would be to lower your expectations. Man, what a bummer that concept is. I know.

It wasn't supposed to be like this. It was supposed to just keep going along the way it always had, things getting better with each generation. Kids would live better than their parents, and so on. It's not working out that way though. Starting in 2008 we found ourselves in the midst of an historic, dramatic, jarring, unsettling shift like we've never experienced before. Many economic indicators are suggesting that we are in a depression and it will be a long and protracted one. That means that the solution to your financial challenges won't come easily.

The reason it's going to take a long time to correct this situation is that we put off the inevitable for so long. It's like this story from nature: For many decades people who managed large forested areas had a goal of fire suppression. They believed that as soon as a forest fire started it had to be extinguished quickly before it spread. On the surface this approach made sense, but what it overlooked was that during each year with no fire more combustible material built up on the forest floor. Over decades all those dead trees and leaves and branches and pine needles got deeper and thicker. When a forest fire started it was much harder to control because there was so much fuel for the fire. In fact now some forest managers actually set controlled fires to try and eliminate or minimize this problem.

A forest fire is a terrible thing, but it's part of the natural cycle of destruction and rebirth. Dead and old material is burned off. Seeds in the soil germinate, some only after intense heat, new growth begins, and the forest regenerates itself. The business cycle is like a forest, and forest fires are like the recessions or economic downturns that are an inevitable part of business. What's happened over the last decades though is that those in charge of the economy have adopted an out-of-date 100-year-old forest management model as their guide. Alan Greenspan (the U.S. Federal Reserve Chair) in particular felt it was his duty to avoid any economic hardship and keep the good times rolling. When there were indicators that the economy was about to go into a natural cycle of destruction or recession he used every means at his disposal to prevent or minimize it and get the economy growing again. So what we're experiencing today is the result of 20 or 30 years of forest-fire suppression unleashed on the economy, and the wildfire that's now burning through the economy with all that combustible material is going to burn very hot for a very long time. The destruction will be far greater than if we'd just let the smaller fires burn themselves out earlier. We're paying the price for an extended period of good economic times that were artificially created.

What came from the prolonged good economic times was that consumers started behaving in inappropriate ways in the context of standard economic theory. They began spending as if there would never be a rainy day and assuming that things like house prices would always go up. These things are not realistic and the inevitable downturn has caught many of us by surprise. But economic cycles, like nature, have a way of leveling the playing field and getting even. If you build your house in a flood plain, sooner or later it's going to be surrounded by water. Going into this downturn North Americans had little or no savings. In fact many had negative savings, owing more than their net worth.

The good news is that many of us are back to saving. We realize now that rainy days do come and that we'd better prepare. And that's the theme of this chapter. It's time to change your behavior and realize that things are not going back to the way they were any time soon so you need to reduce your expectations.

I am going to take a different approach than that of many of the most popular financial advisors you see in the media today. I'm going to suggest a radically different approach. I watch and read advice to people on how to deal with their financial problems and while the goal seems to be to get people back to living within their means there is still the assumption that soon things will straighten themselves out and we can all get back on the spend-and-consume bandwagon. I think we need a much more radical approach. Either those advisors are not being honest or their presumptions are incorrect. Things may not ever be going back to the way there were. Remember these same advisors are the ones who didn't give you a heads-up on what was coming. I don't think they believed it was possible for things to get so bad so fast.

Let me assure you that there have been many voices predicting this economic mess for many years. They predicted its inevitability and they predicted its severity. And now they're predicting its longevity and they're raising the possibility that even after a long period of adjustment things may never return to the way they were. That is something I sincerely believe and my advice is going to be based on that. I simply don't think you can trust prognosticators who were caught completely off guard by the severity of this economic mess. It wasn't on their radar, so I think you have to ignore any advice they're giving about how or when things are going to recover and return to normal. It may not happen. The new normal is going to be one of much less economic activity and much less money to go around. So you need to start preparing mentally and financially for this new reality. And if I've got it wrong and things do get back on track, so much the better. What I'm recommending won't be detrimental to your financial health in any way. In fact, if the economy starts rolling along again as it has in the past, you'll just be in a much better position to enjoy some of the benefits that you didn't enjoy when you were so cash-strapped.

One of the best ways to prepare for an extended economic downturn is to get your financial house in order, and for most of us that means paying off debt. And our single biggest debt is usually our mortgage. During the housing bubble in the early part of the new millennium, North Americans bought bigger and more expensive homes and increased their mortgages accordingly. In a rising market this is a sound financial strategy. Unfortunately many North Americans made the assumption the market would continue going up forever, and its sudden and dramatic reversal has created serious consequences for many. Millions of Americans found themselves under water on their mortgages, meaning they owed more on their mortgages than their homes were worth.

That's the bad news. The good news is that the majority of mortgage holders are not under water, and have an opportunity to take a huge step towards financial independence by whittling away at that debt. You need to pay off your mortgage. Soon. You need to walk away from the financial wisdom of the day which says taking 25 or 30 years to pay your mortgage is a good strategy. It's not. You need to pay off your mortgage and you need to do it as soon as you can. The best way to keep the wolf away from your door is to get that mortgage off your back. Being mortgage-free opens up a world of opportunity. Suddenly decisions that involve a reasonable degree of risk don't result in your losing your home.

I understand what your response will be. Sure, it's easy for you to just say "pay off your mortgage" as if it's easy, but it's not. It's a lot of money and doing it early is going to be a real challenge. I understand that. It is a lot of money and it is going to be a challenge. But you're going to have to change your behavior and do it. You're going to have to focus on this goal like a laser scope on a target and commit everything you can to doing it. You're going to have to alter your behavior from being a consumer to being a saver. You're going to have to stay out of malls and stores. You're going to have to start making do with what you have, and when you really need something you're going to have to find places to borrow it or buy it used. You have to make paying off your mortgage a holy crusade that takes over your life. You're going to be so focused that you start picking up all those pennies and dimes you used to walk past on the sidewalk and putting them towards your mortgage. Americans take great pride in their ability to pull together and attack a challenge, whether it's winning World War II or putting a man on the moon. You have to make that same commitment to paying off your mortgage. This is what a sensible prepper does.

For the last 15 years I've been mortgage-free and that allowed us to uproot our family from suburbia where we had an electronic publishing business and move three hours away to the woods. It also allowed us to evolve the business from doing work for corporate customers to publishing our own books about renewable energy and sustainability. It's unlikely we would have had the confidence to undertake such a journey if we hadn't been free from the constraints of a mortgage.

We also made a decision to purchase only a rural property we could pay cash for. We were not going to take on another mortgage after paying our old one off. It's too enabling a feeling to be free of the obligations of a mortgage to ever go back to those days.

To pay off your mortgage you have to structure it correctly and then you need to be committed to the task. Twenty years ago it was time for us to renegotiate our mortgage. Our existing bank showed us what the new monthly payments would be. A second bank did the same. Then I went to meet with the local trust company. They showed us what the monthly payments would be. Then the representative asked if we'd be interested in paying weekly. Our monthly payments were going to be about $800. I suggested that we did not have the financial wherewithal to pay $800 each week. She replied that the net amount would be the same, $800 per month, but that $200 payments would be taken out of our account weekly. Then she swung her computer screen around and showed us that if we paid weekly we would save approximately $35,000 over the 20-year life of our $66,000 mortgage.

I was blown away! First at how much money I would save, and second at how even though my current bank offered weekly mortgage payments, they hadn't mentioned them to me. Why would they if they would lose a huge chunk of profit? They really didn't have my best interests at heart, but it is a capitalist system after all.

I asked the customer rep why everyone wouldn't pay their mortgage this way? I assumed that most people weren't aware that this was an option, but she said that most people's finances didn't allow for it. They were paid monthly or biweekly and their budgeting didn't allow weekly withdrawals for the mortgage. This is bad financial planning on a number of levels. Mostly it's bad because people live so close to the edge in terms of money in their accounts. This is the wrong way to manage your money. You need to have a slush fund amount in your account that allows you to take advantage of opportunities like this.

So **Rule #1** in terms of money is to pay down debt. To do that, ensure you structure your mortgage to allow the greatest flexibility and fastest payback with the minimum of interest. This can include making the term shorter rather than longer. As the housing market was peaking and financial institutions needed more cannon fodder, they began offering longer-term mortgages of up to 30 years and beyond. This made it easier for some people who couldn't afford a home to suddenly own one. This is too long a term for a mortgage. Make your term shorter to force yourself to pay it down faster.

Shortly after we'd switched our mortgage to the trust company that was willing to let us know about weekly payments, we also looked at some savings we had. The Canadian government at the time gave checks to parents called a "Baby Bonus," which was designed to ensure that children had food and clothes and basic necessities. Under the concept of "universality," every Canadian child received these payments, regardless of the parents' income. Because we were frugal we were able to bank these checks. We called the account the "Kids' "College Fund" account. Eventually it grew to about $9,900.

This was about the time we were starting to feel a desire to move out of the suburbs, and we knew the key to our financial independence would be paying off our mortgage. So we took the $9,900 and put it towards the principal of our mortgage. Our new mortgage had a feature where once a year on the anniversary date of the mortgage we could pay off up to 15% of the principal. So that $9,900 reduced our principal from about $65,000 to $55,000. That was a pretty great feeling, but it was tempered by the realization that this would probably be the only time we could do this.

As the year went on though, we decided to give it a shot again. We took a percentage of every one of our paychecks and put it into our "Five-Year Plan — Pay Off the Mortgage" account. This had previously been the "College Fund." We scrimped and we saved and we put off buying things. We wanted to take the kids to Disneyworld, but that was $3,000 we could put towards the mortgage. We needed a new bed, but flipping the mattress over would have to do for another year. The rust had eaten away at the floor under the driver's side of the car to the point where you could see the road whisking by, sort of like a Flintstone-mobile, but a new car could wait.

Lo and behold, the next year we were able to do it again, and we put $9,900 towards the principal on the anniversary date of the mortgage. I should point out that we were by no means well off in terms of income. The median family income at the time was about $50,000, meaning half the families in the country had incomes below that figure and half had incomes above that figure. We were always comfortably in the lower half. In fact, as a result of the challenges of running our own small business there were some years when our income dropped dangerously close to what in Canada was deemed the poverty line. Canada has a very generous definition of what this income level is, but I share this to show you that you don't need a high income to become financially independent. You need to be frugal and you need to get out and stay out of debt. But you have to be solely focused on this one goal. A weekly trip to the mall will not help you in this cause.

Each year as the car got older and older and we wanted a new one, we held off. Eventually whenever we took a long trip we'd rent a car. Owning and operating a car costs many thousands of dollars a year in taxes, insurance, maintenance, and repairs. We took collision insurance off the car because it was so old that if we had been involved in an accident they wouldn't have given us much anyway. But it was paid off and we elected to save a huge amount of money and keep driving that older car and put the savings towards the mortgage. Keeping the beater on the road and renting a car for longer trips that required a more dependable car saved us thousands of dollars.

For three more years we continued in our single-minded mission of paying off our mortgage. Snowsuits got used a third year when they were probably tighter than the kids would have liked, we came out at the winning end of our Automobile Club membership as we made greater use of tow trucks and services than someone with a newer vehicle, and vacations were canoe trips in provincial parks.

The other brilliant part of paying off a chunk of principal on your mortgage periodically is that as you continue with your regularly monthly or weekly payments a greater percentage is going towards the principal. Mortgages are front-end loaded, so that in the early years of the mortgage you are paying mostly interest and in later years you are paying mostly principal. So what you are doing with these regular principal payments is moving up the time continuum so that a larger percentage of each payment goes towards the principal.

Let's say you had a mortgage of $240,000 with a term of 30 years and an interest rate on the loan of 6%. The national average is that people sell their home every 7 years. If you did this you would owe about $216,000 on the principal after having paid $120,000 in mortgage payments. Of that amount only $24,000 would have gone towards the principal. You would have paid $96,000 in interest.

So **Rule #2** is to have a mortgage that lets you pay off a percentage of the principal periodically. Have this option and use it!

On the 5th anniversary of our mortgage we were able to apply one last check and pay off our mortgage. Of all the feelings in the world, there are few that rival the feeling of leaving your bank without a mortgage. We photocopied our mortgage and had a ceremonial burning in the fireplace that night. (We thought it was a good idea to hang onto the original just in case.)

It was as though a huge weight had been lifted from our shoulders or the storm clouds had left our home and the sun had come out. Suddenly anything was possible. And it's time you experienced this same feeling of freedom. It's going to require sacrifice though. It's going to mean keeping that downhill ski equipment a couple of years longer than you'd like. Heck, it means not downhill skiing at all. It means heading to the local ski swap or reuse centre, picking up a pair of used cross country skis, and finding a forest or trail near your home to ski on for free. It's cheaper and better cardiovascular exercise.

It means getting used to the fact that the vehicle you own will no longer be classed as "late model." It means that the next time your starter motor goes, rather than taking it to the dealer you're going to find a local mechanic who can put in a rebuilt starter. Or better yet it means asking your neighbor who's always out working on cars to coach you on installing a starter motor. I find most people love sharing knowledge. You can do it, and you'll save hundreds of dollars by buying a used starter from a wrecker and putting it in yourself. In the past you've traded your labor for an income that you used to pay someone else do tasks like this. Those days are over. You need to keep that income yourself and save the money by learning a new skill set.

It means saving every penny you can and putting it into the "Five-Year Plan — Pay Off the Mortgage — Move to the Dream House in the Woods" account. You need a separate account for this and you need to celebrate every time the balance goes up. Saving money in our consumer-obsessed culture is a huge accomplishment and you should be proud every time you're able to do it. But don't celebrate with a trip to the local roadhouse and end up $75 poorer by the end of the evening. It's time you learned to make pizza at home and indulge in a six-pack of beer, preferably beer you brewed yourself to save money!

As a consumer you've learned to celebrate earning an income by coming home from a shopping expedition with a bunch of shiny plastic bags filled with "stuff" to show what you've accomplished. When you don't buy "stuff" you don't end up with a little prize. You have done something much harder and far more celebration-worthy, but it comes in under the radar. So start a ritual. Every time the "Five-Year Plan" account goes up by $500, it's pizza and beer night! Every time it goes up by $1,000 it's off to the local second-hand bookstore with $10 to spend on anything you want! Every time it goes up $5,000, well, the sky's the limit! It's rent-a-new-movie night. This is a big deal because up until now you've had to wait until it hits the "Two for Tuesday Night" cheap shelf. Go ahead! Spend $5. See it as soon as it's released! You've earned it!

We've all seen those huge thermometers that hospitals and public institutions use to track fund-raising progress. It's time to make one for your fridge. Set the goal. It might be $20,000 that you want to raise this year to put towards your mortgage on its anniversary. You should transfer $1,000 of each paycheck to that account, and mark your progress each payday. Look at it every time you open the fridge. This helps you stay focused on the goal. It also gets you thinking: All right, if I can keep putting in $1,000 a month, we're still $8,000 short of the goal. Where's it going to come from? What about that guitar I don't play any more, the one Rick Nielsen of Cheap Trick signed? Wonder what I could get for it on eBay? And those *Mad* magazines from the 60s gathering dust in the basement? Maybe it's time to see what they're worth. And Tom down the road needs some help installing hardwood floors on the weekend. Maybe it's time I started helping him out to earn some extra cash.

If you've got a mortgage you want to pay off, you've got to be obsessive about it and a thermometer on the fridge is going to keep it in your face. If you've been having back problems and you've convinced yourself that a $5,000 hot tub is the solution, being constantly reminded about how much that purchase would set you back from your goal may finally get you down to the library to borrow some yoga tapes and get you doing some stretching exercises first.

If you're reading this and you don't yet own a home, all the better. This housing market represents a tremendous opportunity for you. The glut of homes built during the housing boom will mean that there are more places to rent, which should drive rents down. It also means that when you've saved up enough to buy a home it will finally be affordable. I am shocked at the prices of houses. Part of this is the economic environment created by Alan Greenspan and policies such as mortgage deductibility in the U.S., which encourages home ownership. One of the other factors is how large houses have become. The average American home has doubled since the 1950s. In 1950 it was 983 sq. ft., in 1970 it was 1,500 sq. ft. and by 2004 it was 2,349 sq. ft. We raised our daughters in an 800 sq. ft. bungalow that had two bedrooms and one bathroom. They shared a bedroom and in their twenties they're still the best of friends.

So if you're hoping to someday own a home, don't rush it. When Ben Bernanke took over as Alan Greenspan's replacement as Chair of the Federal Reserve he was on record as saying we had never experienced a cross-country decline in house prices. Because it had never happened, he couldn't conceptualize it. But it has come to pass with a vengeance. House prices went into free fall across the country. So the longer you can wait to buy a house the better. You should still save your money with the same zealousness as someone trying to pay down a mortgage; you're just increasing the size of your down payment, which will mean less principal on your mortgage and therefore less interest to pay in the long run. The more you put down, the smaller the principal amount, and the faster you'll be able to pay it off.

If you have a mortgage and you have an income your goal is simple. Pay off your mortgage, first and foremost. If you don't own a home and are looking at a market where house prices are declining, it's better to be renting that asset for now and then owning it when the market hits bottom and begins to improve.

The "Thriving During Challenging Times" workshops I have been giving for a number of years have evolved from the renewable energy independence workshops I began doing a decade ago. One of my contacts at a college where I run the workshop saw my description of it and noticed that it included a financial component. She called to question whether I was a certified financial planner. I assured her I was not. I also explained that my financial advice was pretty basic. In fact the whole financial theme of my workshop came down to one PowerPoint message which I emblazoned across the screen in the biggest font that would fit. It was simple. The key to financial independence is to:

STOP BUYING STUFF!

Seems pretty basic, but these are three very loaded words which contradict every message we receive through the mainstream media. Americans are consumers and personal consumption accounts for 70% of gross domestic product. That's a staggering amount and it's what has been keeping economies like China's humming along, trying to keep up with the endless demand for "stuff" for Americans to buy. Acquiring stuff you'd think would make us happy, but it doesn't.

Now I know what you're going to say next. "But you said I should be buying a solar thermal system to heat my hot water and solar panels for electricity." You're right. You caught me. So let me rephrase it. "Stop buying stuff—that doesn't make you more independent." Solar panels and gardening tools help you move towards independence. Designer shoes, sports memorabilia, a 1960 Ford Mustang, a riding lawn mower, a ______ (fill in the blank with about a billion useless things we all buy), don't make you independent. They just fill your life up with clutter. We now have TV shows on how to get rid of stuff.

At the same time that American houses are getting bigger and bigger we have this whole industry of storage units where people take even more stuff to store off-site. How much "stuff" can one family own? This is part of the reason we're in the mess we're in. It used to be that people saved some money. The bank took that money and lent it out to other people in the form of mortgages so they could own a home. Now banks often don't have enough savings on hand to do that and have to look elsewhere for money to lend. And some of the money that comes in to support our North American lifestyle comes from China, where people still save money.

So you need to return to those days when people saved money. You've now got at least two new bank accounts open for specific goals to save towards. With separate passbooks at separate financial institutions that you can ride your bike to conveniently. You have the "Solar Power" account where you're putting all your energy savings to buy solar panels for your roof. You've also now got a "Five-Year Plan/Pay Off the Mortgage" account where you're putting every other penny you can save.

Credit Card Debt

One of the things that are going to keep you from achieving those savings goals are those magical pieces of plastic in your wallet. Credit cards are incredibly convenient and a great invention. But they're like many other inventions. Morphine is an incredible painkiller, but if you use it too much and for too long you get really addicted to it. In the early stages of its use it serves an amazing function: preventing extreme pain. Eventually, though, if you don't wean yourself from it it becomes as big a problem as what you needed to take it for in the beginning and you become addicted.

Credit card debt is an absolute no-no and you need to eliminate any credit card debt you have. It's much worse than mortgage debt because with a house once you eliminate that debt you have an asset that has great value. You can live in a house. It keeps you warm and comfortable and for many decades it maintains its value. The stuff you purchase with your credit card generally will not maintain its value and you've purchased a depreciating asset. If you've been carrying a balance on that credit card to pay for these things that lose value you're burning the candle at both ends.

Paying credit card interest is to your household financial health what smoking is to your personal health. It's bad and there is no upside. Ultimately it's going to severely weaken your ability to thrive financially. So you've got to pay it off. I'm hoping you can do this fairly easily and quickly. I'm hoping you have a couple of credit cards and a couple of thousand dollars of debt that you can focus on and get rid of even before you focus on your mortgage. My concern is the trend that I've seen recently on several TV shows that focus on people with severe debt problems. Sometimes they have 10 or 15 credit cards and they have tens of thousands of dollars' worth of debt. If that's the case you have to take drastic action. You have to go on a crash diet and give up everything but bread and potatoes until you pay them off. Sell the car, sell the motorcycle, sell anything and everything and don't buy anything until all your credit card debt is paid off.

If you were hoping for some magic bullet to make this disappear, I'm afraid I don't have one. A huge balance owing on multiple credit cards is going to take some real time and commitment to get rid of. And you have to because the interest you are paying on credit card balances is very high. It is much higher than your mortgage interest and higher than a line of credit which is secured against assets and therefore usually offers a better interest rate.

Annual interest rates range from 8.99%, which you would pay if you had an excellent credit rating, to up to 30% for someone without a good credit rating.... OUCH! Let's take the example of purchasing an item for $1,000 on your credit card and only paying a small or minimum amount each month. If you had a good credit rating and were able to get a credit card with a low annual interest rate of 10% and you paid $20 per month, you would end up paying $300 in interest and it would take you 5 1/2 years to pay it off. If, however, you did not have a good credit rating and your credit card had an annual interest rate of 20%, that same $1,000 purchase at $20 per month would take you 9 years to pay off and you would end up paying $1,171 in interest! If you were paying a rate as high as 25% and you only paid $20 per month you would never pay the balance off. After 10 years you would have paid several thousand dollars in interest and still owe the original $1,000.

So if that $1,000 item costs you $2,171 ($1,000 + $1,171 interest) does this sound like a wise thing to do? Wouldn't it be easier to save the $1,000 and purchase the item for cash? Credit cards are a trap, just like the quicksand you used to see people fall into in old movies. Don't get caught in the quicksand or credit-card trap. So stop buying stuff you cannot afford!

Under their "Consumer Tools" the National Foundation for Credit Counseling (www.nfcc.org) has some excellent calculators that allow you to plug in your account balances and interest rate to get an idea of how much interest you'll end up paying. If you have debt right now go to this website and get a handle on the reality of what it's costing you.

You can see that paying $300 interest on a $1,000 purchase is a lot of money. Paying $1,171 in interest on that same purchase is insane. It's bad money management. You simply have to stop using credit cards in this way. If you need to buy a new solar panel to add to your system, you need to save the money to do it and pay cash for it. If you've researched solar panels and have a pretty good idea of what they're worth and you happen to be at a renewable energy fair and a dealer is putting away her booth at the end of the show and offers you the panel at an exceptional price to save having to truck it back to the shop and you don't have cash on you, then it's all right to use your credit card. You have less than a month to come up with the money to pay off that amount on your credit card bill when you get it. This is the way you have to think about credit cards if you have them. They are a temporary tool, not part of a sound financial strategy.

If right now you have multiple credit cards with balances owing, pick the card with the highest interest rate and pay it off first. Then take out those scissors and cut it up. Celebrate it. Have a pizza party. A homemade pizza party. Then take the card with the next highest interest rate and pay it off. Many financial advisors today say that once you pay it off you should put it away and just not use it. This is because your credit rating will be better if you have lots of credit cards. They suggest you just use it once in awhile to keep it active but pay it off. I completely disagree. Credit is a drug and if you've been paying interest on a credit card you're a junkie. So stay away from the dealer and cut the card up.

If you have multiple credit cards with balances some credit counseling services suggest you try and get a consolidation loan to pay off the credit card companies and roll all that debt into one big pot. Then you can begin paying off just that loan. The interest rate will be sane and you'll get out of that spiraling trap of compounding interest on interest. If you do have multiple cards with balances it may be worth asking a credit counselor to take a look at where you are and provide some feedback. The National Foundation for Credit Counseling is a non-for-profit resource for this and can be reached at 800-388-2227 or nfcc.org.

The biggest factor in success with using an outside support person is that often they'll be able to step back and take a big-picture look at your situation to make sense of it. When you're directly immersed in the mess sometimes it's difficult to be objective about the solution. It's like having your mom or dad tell you to do something when you're a kid. Sure you're an adult now, but if someone behind a desk in an office is what you need to get your credit card debt in line, then by all means, get one on your side.

Many years ago I went to an investment counselor to invest in the stock market. He went through an inventory of our assets (none, we rented an apartment and had a car loan) and our incomes, which were moderate at that time. Then he asked how much we had in savings. "Savings?" I asked. He said, "You make that much money and have no savings? Get out of my office and don't come back until you do!" I think I was about 25 at the time but felt as though I were 7 and my father had just yelled at me for setting fire to my plastic soldiers in the sandbox. I was mortified. From that day on Michelle and I started saving money, and we've been savers ever since. That's the advantage of getting advice from an outside person. Often you'll do a better job of getting your house in order for them than you will for yourself. I'm sure there are many books written about what in our makeup causes humans to behave like this. I'm guessing it goes back to pleasing your parents, but that is another book. But who cares. Once I started saving it was like: "Ken (my financial advisor) will be so proud of me!" And sure enough six months later we went back and he was impressed and decided to take us on as clients.

If you have manageable credit card debt, say $2,000 to $5,000 on a couple of cards, you should just put your nose to the grindstone and pay them off quickly. With the interest rate you're paying on a credit card it has to go first, and it has to go soon. You can't start making any plans about independence if your income is being sucked into the vortex of credit card interest.

I created the analogy of credit card debt being like smoking and nicotine to your financial health, but I think nicotine may be a little tame. Credit cards themselves are like any of those addictive drugs like cocaine that we hear so much about. Why? Because they create an illusion. Cocaine creates an illusion of well-being. Credit cards create an illusion of financial well-being that isn't reality. They allow us to walk into a mall and think that, regardless of our income, anything is possible (as long as it's less than our credit limit). It creates the illusion of infinite abundance when for just about everyone money is finite. It has limits based on our income and assets. Credit cards let you break free of the bounds of gravity imposed by your income and soar to unimaginable heights. But just like Icarus and his wax wings, when we soar too high we end up being brought back down to earth, often with a thud.

Until recently one of the advantages of credit cards was the convenience of being able to substitute them for cash and not having to carry so much cash with you. But debit cards have changed that. Debit cards are much more fiscally responsible because they are like cash. The retailer you're completing the purchase with simply removes the money from your account electronically. If the money is not there, then you're going to get turned down. A debit card has the convenience of a credit card but with the reality check built in. If you don't have the money in your account, don't buy it. So scrapping your credit cards and going to a debit card makes sense, although if you know you have the resources to pay for the item, it's nice to have a credit card to fall back on. Having one credit card that you only use for emergencies and always, always ALWAYS pay off completely each month is also not such a bad idea.

I have one credit card. I use it because it's convenient and because I get reward points from a hardware retailer I like who also sells gas. As much as I'm trying to reduce my carbon footprint, I still drive and I still buy gas. This retailer is rewarding me so I buy it there. I don't buy more than I need in order to earn points. I just earn points on what I'd be buying anyway. And since I get to spend my points in a hardware store, I can always find stuff that helps me in my quest for independence, whether it's tools to fix things or gardening supplies. I know what you're saying. I'm part of the consumer culture! I'm getting rewarded for being a consumer! I'm a hypocrite! OK, you got me. But I still live in a capitalist society and it sells a lot of amazing products that make my life easier and allow me to be independent. Gas is gas. It's just a commodity so I'm putting gas in my tank and this retailer is going to give me some points I can use towards a new cultivator for the garden. I'll go there.

Line of Credit

A line of credit is a vehicle that allows you to borrow money against a fixed asset like a home. As the housing market was on its way ever upward early in the millennium many people got very familiar with lines of credit. With the value of homes increasing, a bank or financial institution was often willing to loan you money to purchase things because the loan was secured against the house. The mortgage holder would be first in line to get paid if you defaulted, but the line of credit was often issued by the same institution, so it was just another way for the bank to make more money. You paid them interest so that you could have something immediately rather than waiting until you could afford it. This in itself is not a bad thing.

Our family has a holiday tradition of watching *It's a Wonderful Life* on Christmas Eve. This is an outstanding movie on so many levels. In one scene Jimmy Stewart's character George Bailey is ranting to the evil banker Mr. Potter about how his father's Savings and Loan Company had loaned money to people the bank had turned down for mortgages. Potter suggests they shouldn't own a home until they can pay cash. George Bailey disagrees and suggests it's really not such a bad idea for a family to have the money to purchase a home when they need it while their families are young rather than wait for 25 years while they save and buy it when the kids have left home. It gives them dignity and pride in where they live and a sound financial asset at the end of the process. "Well, is it too much to have them work and pay and live and die in a couple of decent rooms and a bath? Anyway, my father didn't think so."

I agree with George Bailey. Using a mortgage to buy a home for a family is a wonderful part of the financial system. It's an admirable thing to allow a young family to take pride in home ownership and end up with a sound financial asset. Where we might be creating a problem though is with lines of credit. As housing prices kept going up, lines of credit allowed forever-increasing expectations on the part of homeowners in terms of what they could have. Want a boat? A nice big boat? No problem, use your line of credit. Want a high-end SUV and can't afford it? Forget about it. Use the line of credit. Always wanted a designer kitchen just like the ones on those TV shows? It's as easy as writing a line of credit check. People were able to use their homes as bottomless ATMs and just keep withdrawing cash that was appearing out of nowhere as their house magically went up in value. It was "easy come" until we watched the devastating impact of "easy go."

Somehow our instant-gratification culture forgot that sometimes good things are worth waiting for. It's not necessarily realistic for a young couple to own a massive home immaculately furnished and with two huge SUVs in the driveway. Remember the olden days? The 50s and 60s and even the 70s where you rented until you could afford a half-decent down payment? And when you did buy that first tiny fixer upper, your voice echoed around inside because there was so little furniture to absorb the sound? Remember the crappy linoleum floor and tacky countertop and stained sink? People who are now in their 50s grew up in those houses and really didn't mind it. There weren't as many high-tech gifts under the Christmas tree but a new tricycle and a plastic gun could bring such joy to your life. Remember how that tree with the rope swing could amuse you for hours? How is that possible? There was no monstrous wood and plastic climbing structure with a built-in play house and climbing wall, and yet somehow most of us were pretty happy as kids.

So the line of credit has to go as well. Or at least using it for frivolous purchases has to stop. A solar domestic hot water system purchased on your line of credit when you know it will be paid off in three to five years is probably not such a bad idea. You get hot water from the sun and as long as you pay down that line of credit you end up with an asset that's going to reduce your carbon footprint and save you buying natural gas for as long as it works. That's a good use of a line of credit. A granite countertop will not help you achieve financial independence. A solar hot water system will. Some government programs require that you do energy-efficiency and renewable-energy upgrades within a fixed time period after an energy audit. If you don't have the cash to complete everything before the time expires but there is a significant rebate if you get the system installed, use the line of credit. Get the solar domestic hot water panel installed and as soon as that government rebate check arrives in the mail put it towards the line of credit. This is an example of a line of credit being a useful tool in your quest for independence.

So if you have a line of credit with nothing on it that's a good thing. Just leave it there. Don't use it unless you can really cost-justify the purchase of an asset. If you have a balance on it, pay it off. Like credit card debt the interest rate will be higher than your mortgage, so pay it down before you get cracking on your mortgage. The order should be to pay off your credit card debt first, your line of credit next, and then your mortgage. Some people may find they have a great rate on their line of credit and may want to use it to pay off all those credit cards. Just remember that you then have to double your efforts to pay off that line of credit. It's secured against your house. If you can't pay off that line of credit, remember that the asset that secures it is the one you call home. So get serious and pay it off.

Forced Savings

Many of use are familiar with retirement plans where money is put into our retirement fund directly from our paycheck, even before we see it. This is forced savings. While you could get your paycheck and do this yourself, for many a plan like this ensures that the money gets where you want it to go. Well now that you've made the commitment to be more financially independent you have to take it to the next level. This means creating a forced savings account. These are after-tax dollars, which means the money you get from your paycheck. It ends up in your bank account and there's a huge temptation to do something else with it. Buy a big-screen TV, take a trip to Vegas, get that motorcycle you've been thinking about. Forget about it. They're not going to happen. You need to have discipline and immediately put that money into your forced savings account.

You can prioritize and decide what the account is for. It can be the "Solar Power" account, the "Five-Year Move to the Country Plan" account, or your "Emergency Fund" account. Whatever it is, it has to become habit that you put money into this account first. It's the "Pay Yourself First" budget plan. Before you pay the satellite TV bill, pay yourself. If by the time that satellite TV bill is due you don't have the money, you're going to have make a choice. Are you going to put it on a credit card? We now know you're not. Are you going to not pay it? That's not the best option since they won't be happy with you. Or are you going to call them, cancel the service, pay off the remaining balance, and hook up that antenna that your grandparents have stored in their backyard? Oh, you won't have the number of channels or the picture quality, but you will get your TV reception for free. This is what we're talking about. You've got to make hard choices if you're serious about becoming financially independent. There are no easy tricks.

Emergency Funds

The first forced saving fund you'll be setting up is your "Emergency Fund," which is designed to provide your family with a cushion against the loss of an income. The amount varies and it's actually an irrelevant number because there are so many variables. Some financial advisors use eight months as the number. Some say three months. How about a year? You should calculate a monthly budget and once you know how much money you need to cover your expenses for a month, from the mortgage payment to food, utilities and insurance, multiply that by eight months and that's how much of an emergency fund you should have.

Obviously the lower your monthly expenses the smaller your fund can be, which is another great reason to reduce your levels of consumption. The smaller your home and the more affordable your insurance and expenses the easier time you're going to have in the event that someone in the house loses their job. So by all means have an eight-month emergency fund. Have a year-long emergency fund. Ultimately, though, what you're working towards is the day when the loss of a job won't be as big a deal because you'll have your mortgage paid off, you'll not have debt, and your living expenses will be minimal. Installing a solar hot water system will mean less natural gas or electricity for hot water. By installing solar electric panels you won't require so much grid power. The goal of paying down debt has the added bonus of increasing your financial independence and lessening the impact of job loss.

So pick a number—6 months, 8 months, 12 months, whatever works for you—and set that money aside in a government-insured savings account. This shouldn't be considered a financial asset, just a safe and secure, easily accessible savings account. If you can earn some interest on it great. But growing this asset is not the goal. Keeping it safe for when you need it is paramount. As you're developing these forced savings accounts—for emergencies, renewable energy equipment, your yearly mortgage principal payment, and your five-year plan—give some thought to having them at different financial institutions. This spreads the risk of a bank failure. It also increases the pleasure you'll start taking in saving money. Each payday involves a trip to two or three different institutions on your bike to make a deposit into a savings account. When you leave you can look at the passbook and see how you're doing. People like to keep score and this is an excellent way to do it. If you just have one savings account you use for multiple purposes it won't have the same impact and it becomes difficult to keep track of how much of the total is for each goal. Having separate bank accounts in different banks is just way more fun.

Monthly Budget

I've already mentioned the necessity of a monthly budget to help you calculate the amount you'll need for your income loss emergency fund, but it's worth repeating just how important it is to actually have a budget. Most people have computers, so there's no excuse. You don't even need a computer for that matter, although there are lots of inexpensive home budget programs if you don't want to just make your own on a spreadsheet. Sometimes doing it on a computer makes it more inviting and therefore more likely to get done.

Budgets serve two purposes. The first is knowing what you need to live on. The second is to expose just where you're spending your money. I think most people are surprised when they actually see where their money goes. Sometimes you think you have an idea of how much you spend on things like groceries, but then when you see the actual amount you spend you're flabbergasted. That's one of the beauties of a budget. Having accurate information about your finances is an excellent tool for everything else we've discussed in the book. You won't know whether you can afford a new wind turbine unless you know where you're at, and using the seat-of-your-pants method where you guess whether you have the money or not doesn't work anymore. Credit is tighter than it used to be and you've seen the necessity of living within your means, so putting it on the credit card doesn't work. That purchase will have to wait.

So set up a budget and stick to it. First set up all your income sources, then expenses that are withdrawn from your account monthly, like insurance and car payments.

Keep all your receipts and start inputting them into the budget. It's crucial that you track everything from your morning coffee to magazines you pick up for the subway ride home. Two $5 coffees at work is $50 a week or $2,500 a year. It doesn't seem like a lot each time, but two years of making your own will almost buy you a solar domestic hot water heater. Then half your hot water is free every year and instead of sending that money to the gas company you can use it to pay off your mortgage. It's all about having good information. You can't make good decisions without information about money in and money out.

After a few months sit and down and analyze your budget. What's going well? Where could you improve? Whoops, we spent $400 at restaurants last month. We need to cut that in half. Someone spent way too much money on clothes. Time to find some local second-hand shops and reduce that. We spent THAT much on food? All right, why did we spend so much? Too much of it was processed and ready to eat. So we need to start making more meals from scratch. There will be an infinite number of ways to reduce expenses once you actually know how much you're spending.

Investing

Oh great, now we get to talk about investing! This is the best part. This is the part most financial books talk about. In fact, there are tons of books on just this topic! So I think I'll leave those books alone and give you an entirely fresh perspective on investing. I think right now what you need to do is forget about investing, forget about growing your assets and looking for a reasonable return on investment, and concentrate on capital preservation. Wow, that sounds complicated! Well not really. I'm just suggesting that rather than trying to make money you focus on not losing it, and in this market that's a real challenge.

The stock market returned to its pre-collapse 2008 levels but one might question how did this happen? Are the basic economic forces actually there or is it a case of the money supply being inflated through 'quantitative easing' (money printing) and with an interest rate of close to zero, do investors have no choice but to put money back in the market to try and earn some return? They certainly can't get a return parking it in a savings account.

With the financial crisis the reality of retirement changed for many of us. It's simply not going to happen for many of us, and for many others retirement is going to happen in a much more diminished way. I hope this doesn't sound too cold-hearted, but I don't see this as a bad thing. I've always found the concept of retirement a little depressing. I guess this comes from some of the things I see seniors doing, like mall walking and sitting on porches looking bored. They just don't seem to have enough stuff to do and can get pretty bitter about it. Perhaps it's because I know of so many people, men in particular, who defined themselves through their work, then retired, and died shortly thereafter. I don't know whether there's any empirical data to back this up, but my first-hand experience tells me that with some people it's the case.

As we all adjust to the new economy, one with fewer jobs and a reduced level of economic activity, money will be harder to come by and sharing accommodation and sharing family responsibilities will become more and more important. Working parents who have had their parents close enough to provide child care have been very fortunate up until now. The necessity for the older members of the family to help with the younger is going to become more important than ever. This is the way families used to operate, and in the new economic order families will often need to return to this method of operation. This is the beauty of the concept of humans organizing themselves into these units we call "families." Sure, lots are dysfunctional, but the concept is still good. Provide a support network, nurture the young, support the elders: it's a pretty sound idea.

Families being closer together is actually a really, really great thing.

Think of how wonderful it will be to value our seniors once again. To have their wisdom and experience close at hand. Think of how much grandchildren will learn spending more than just Christmas and birthdays with their grandparents when everyone's stressed out and behaving badly. And think how much more valued and loved seniors will feel when their contribution to the family is recognized and appreciated. All over the world different generations live under the same roof, and during much of the history of North America economics dictated this lifestyle. We are simply returning to a system of living that we have only moved away from in the last few generations. I am always amazed at how many people I meet whose kids live in Seattle, San Francisco, Boston, or Paris. Cheap and abundant energy has allowed this wholesale dispersion of families. I believe that with the ongoing economic malaise and the end of cheap oil there will be a huge incentive for families to get closer together geographically.

I'm not suggesting that all families will have to move in together. Some people will still be able to retire as before. Some may have to continue working longer than they would have preferred. Some will have to take on part-time jobs to make ends meet. But none of these things is necessarily a bad thing unless you choose to perceive it that way. Having a sense of purpose makes people happy. Contact with others and interacting with younger people helps keep older people mentally engaged. Tending a garden that reduces the family's expenses is something that is going to become really important and it's going to make senior members of the family feel very happy to be making such a huge contribution. Exercise is what keeps people young. Load-bearing exercise is what doctors recommend to fight osteoporosis. Being outside nurturing the garden is a far better way to spend your golden years than watching reruns of *The Golden Girls* all day.

This idea that the concept of retirement as a time of leisure and luxury is approaching its end is not going to sit well with a lot of people. We've really only had one generation where it was the norm. The generation born after the Great Depression was basically the first and only generation that will get to experience retirement as we know it today. Many groups in society worked hard to build up pensions. Autoworkers fought for it, government workers demanded it, and in the 50s and 60s and 70s companies that wanted to keep good employees offered it. But we saw globalization move jobs offshore and the security of full-time and lifelong employment being replaced by contract work and jobs in developing economies. During this economic meltdown all pensions suffered a real setback. Much of the money that had been put aside to pay for pensions was invested in equities and it has suffered huge losses. People who had personally controlled retirement funds were lulled into a sense that stocks would always go up, and since they had never suffered a dramatic drop in stock values they couldn't conceive of anything else. Now reality has set in. What goes up can come down, and sometimes comes down hard.

So what is my advice for investing for your retirement? My advice is to start living in the moment and enjoying yourself today because projecting what the world and the economy are going to look like in 10 or 20 or 30 years is a fool's game. It won't look the way it used to and it won't include limitless growth and opportunity. I suggest you start creating a lifestyle that you can continue to enjoy for a very long time. Living frugally. Reducing your expenses. Growing your own food. Powering your own home. Living within your means. Reducing your demand for "stuff" and therefore your footprint on the planet. We all need to experience "simple abundance" and start to enjoy simple pleasures like harvesting carrots from the soil we have worked hard to condition.

I think you need to first get to a stage where you are debt-free. That has to be your first goal. This is going to take some time if you've been living the North American Dream of spending without limits. Once you get to that stage you need to create a home that is as self-sufficient as possible, from how you heat and cool it, to where you get your electricity, to how you heat your water and where you get your food. A return to the spirit of independence and self-reliance that built this continent and inspired your ancestors to come to this place. This is sensible prepping.

Once you become independent you can start saving money again. That money should be kept in cash or precious metals and I'll discuss this in the next chapter. Some cash you should leave in a bank that you have confidence in. It has to be backed by the Federal Deposit Insurance Company (FDIC) in the U.S. or the Canada Deposit Insurance Corporation (CDIC) in Canada. You need to be critical of those institutions and consider whether or not they can make good on their obligations, in other words whether they can pay you back if the bank fails. This insurance is backed by the Federal Government and you have to decide if you think that's good enough for you. The government has been a very stable and secure institution for a long time, but the economic crisis is putting enormous pressure on governments, which are throwing money at this problem, money that they have to borrow or create.

It's hard for us to conceptualize, but governments do go bankrupt. Governments do default on their loans. The dollars they issue are only as good as the confidence people have in them, and if people lose their confidence then those pieces of paper can become worthless. That is why my advice so far has been for you to have hard assets, things that actually do something. Solar panels make electricity or hot water. They accomplish something, they improve the quality of your life and make you more independent. They are a sound investment. Purchasing shares in a company is a soft asset. You may get a monthly piece of paper showing your ownership in that company, but it is a piece of paper that represents electronic pixels on a computer somewhere. If your statement shows that you own shares in Bear Sterns, the value of your investment went from $80/share in April 2008 to $40 a share in a matter of months, then dropped from $40/share when the market closed on Friday, March 14, 2008 to open at $2/share on the Monday morning after the forced sale to J P Morgan Chase. The same can be said for Lehman Brothers, which went from its 52-week high of $67 to bankruptcy in a matter of months. That is a paper asset and it lost its value as easily as if the paper had blown away in the wind.

Your solar panels, on the other hand, assuming you have them fastened down well and they don't blow away in a hurricane, will keep producing electricity and hot water as long as the sun comes out. The monetary value of that hot water and electricity may change, but the value of what you can accomplish with what they produce, like taking a shower or using a washing machine to clean your clothes, never diminishes. The way electricity replaces human labor is an absolute wonder to experience. We've come to assume that electricity has always been there and always will be, but we cannot take it for granted. It's just too valuable to be without.

The turbulence the world economy and world financial system began experiencing in 2008 may continue for a long time. Harry Dent in *The Great Depression Ahead: How to Prosper in the Crash Following the Greatest Boom in History* suggests it could last until 2017. Others have suggested we are into a long L-shaped depression which is going to continue for decades. If at some point you feel the worst is over and the markets are recovering then choosing to invest in stocks again is your right. You may also want to be more conservative and invest in bonds, particularly government bonds. Treasury Bills have been considered one of the safest investments you can make. You give up the potential of a great return, but you do get security.

I cannot tell you how long the depression will last, where the stock market will go, whether or not governments will default on bonds, and if and when things will turn around. I can tell you I believe that with the number of challenges facing the world today the shiny happy days of never-ending growth in the stock market are over. The easy money was made and now much of it has been lost just as easily. I am not advocating the easy way out. I'm not suggesting there is a simple solution to the mess or a foolproof way for you to make your money grow. I don't think there is. I think you need to change your relationship with money and reduce your expectations in terms of your financial future. This won't get me elected to office, but I think it will help me deal with a much more subdued future. One where people travel less, buy less, eat food grown more locally and support workers in their community rather than 12,000 miles away. This is a good trend for the planet and a good trend for communities. It will make them much more livable.

You may have read this chapter and thought about other things you could be doing financially. You've probably even thought of specific new products on the market that you could use to speed the process of growing your independence and wondered why I haven't mentioned them. Truth is I don't know about them. I don't because I don't need them. I don't have a mortgage, I don't have any credit debt, I have money in my savings account, I have a few bars of silver, so I'm out of the loop when it comes to the wild and crazy world of modern financial products. While you should check them and see if they can help, your financial strategy shouldn't be complicated and need long-winded descriptions. Stop buying stuff and save your money! That's all you need. All the other stuff is fluff.

All the fancy reverse mortgages and financial instruments just distract you from the basic goal of this book. To get independent you need to save more money and spend less. You need to pay off debt and start saving. It's not rocket science and the more complicated the product someone is recommending the more it may be distracting you from the basic goal. It's really simple. Stop buying stuff. Pay down debt. Save money. Period. End of story. This is what makes a sensible prepper.

22 Mediums of Exchange

Once people were able to create money at virtually no expense, no one ever resisted doing it to excess. No paper currency has ever held its value for very long. Most are ruined within a few years. Some take longer.

Some paper currencies are destroyed almost absentmindedly. Others are ruined intentionally. But all go away eventually. By contrast, every gold coin that was ever struck is still valuable today, most have more real value than when they first came out of the mint.

William Bonner, *Empire of Debt*[1]

There was a time when a paper dollar was backed by a valuable asset, a precious metal like gold or silver. Today that's not the case. A dollar bill is a "fiat currency" that has value because a government says it does. It only has value as long as everyone continues to believe it has and continues to have confidence in the government that created it. The U.S. government is now spending money in unprecedented amounts to solve the financial crisis. Its current debt, the amount it has spent over time that it has not paid back, is over $16 trillion dollars. This is how much money the U.S. government owes. It's like a credit card debt. President Obama's first budget of $3.6 trillion added another $1.75 trillion to the total and he has continued to run budget deficits each year while he tries to get the economy working again.

A trillion is a huge number. It's an incomprehensible number. This is what it looks like: $1,000,000,000,000. A million is a big number. A trillion is a million million. It's a thousand billion. So $16 trillion is $16,000,000,000,000. Wow! That's big.

How can it ever be paid back? Well, to be fair I'll give the usual rationalization. Economists say that as the economy grows the government will be able to pay it off. Over time it's hoped that it will be whittled away and become a smaller percentage of the gross domestic product or GDP.

The problem is that on May 28, 2008 Richard W. Fisher, President and CEO of the Dallas Federal Reserve Bank, estimated the obligations of the U.S. to be actually **$99.2 trillion**. Today the U.S. Debt Clock estimates the obligations of the U.S. taxpayer rise to an impossible-to-repay sum of **$123,000,000,000,000.00 ($123 trillion).** These include obligations the Federal Government has for Social Security, Medicare, Medicaid and National Prescription Drug Plan. These are called "unfunded liabilities" because they are money the government is on the hook for but hasn't put into a savings account yet. So as all the baby boomers start retiring and demanding their due from the government, it's going to have to come up with the money somehow.

In the 1970s the debt was about 30% of GDP. In 2012 GDP in the U.S. was about $15.6 trillion with the debt being $16.6 trillion. So debt is about more than 100% of GDP. How long can this continue? No one knows. I guess it can continue as long as people continue to have confidence that the U.S. government will be able to pay the debt back. If it can't, then the value of those dollar bills in your wallet could change radically and quickly. They could have a lot less value. Eventually they could become worthless. Sure that's a worst-case scenario, but sometimes it's a good idea to hope for the best and plan for the worst. The people who say that the U.S. dollar could never lose its value dramatically are often the same ones who said we would never have another depression. They say we learned our lesson the last time and would never let it happen again. And yet in 2008 there we were with the worst worldwide economic crisis since the Great Depression.

One result of an economic crisis like this is that governments try and revive the economy by increasing the money supply, simply cranking up the printing press and pumping more money into the economy. While this sounds logical and easy it's dangerous because it is highly inflationary. When everyone suddenly has more dollars to spend and there hasn't been a proportionate increase in the goods and services people purchase, the people selling things start raising their prices. If more people want to buy things and the demand goes up without the supply also increasing, prices go up; this is what we call inflation. More dollars are chasing fewer goods, so everything costs more. Having more dollars in your wallet isn't necessarily a huge advantage because each dollar buys proportionately less stuff. Increasing the money supply to deal with economic downturns has been a common strategy. Many governments over the last few decades have fought the temptation to do this overtly, but there is a strong possibility it could happen again.

With the negative connotation that inflating the money supply can have, the government is now calling this "quantitative easing," but it's the same old thing—inflating the money supply or increasing liquidity. This really is one of the government's last hopes because it can no longer use interest rates to try and reignite the economy. The interest rate is essentially zero, so it has no room to move. This is exactly the problem Japan has experienced for the last two decades.

This inflation is actually eating away at your dollars, making them worth proportionately less. Some economists like John Williams (www.shadowstats.com) suggest that governments are actually reducing the value of your dollars by not honestly reporting the money supply. In 2006 the U.S. Federal Government stopped reporting the broadest measurement of the U.S. money supply, the "M3." Texas Member of Congress Dr. Ron Paul says that, "M3 is the best description of how quickly the Fed is creating new money and credit. Common sense tells us that a government central bank creating new money out of thin air depreciates the value of each dollar in circulation. Yet this report is no longer available to us and Congress makes no demands to receive it." [2]

Kevin Phillips in his book *Bad Money: Reckless Finance, Failed Politics and the Global Crisis of American Capitalism* analyzes how the methods the government uses to report common economic indicators like the consumer price index (CPI), unemployment, and the money supply all give a false indication of what most of us are experiencing in the economy. In a follow-up article in *Harper's Magazine* in May 2008 he says:

> Readers should ask themselves how much angrier the electorate might be if the media, over the past five years, had been citing 8 percent unemployment (instead of 5 percent), 5 percent inflation (instead of 2 percent), and average annual growth in the 1 percent range (instead of the 3–4 percent range).[3]

What he is suggesting is something that many of us already know, that consumer prices seem to be getting higher than reported. Most mainstream media report "core inflation," which is inflation that does not include food and energy. Apparently they believe that most people using the information have no need to heat and cool their homes, drive to work, or eat.

The chance of a systemic crash is much higher today than it's been for a long time. The chance that inflation will return to the economy in a very aggressive manner is a strong possibility. Let's hope the powers that be can prevent this, but since it's a possibility a sensible prepper should have a plan—a plan to deal with paper dollars having less and less value over time.

Step One – Hard Assets

We've already talked about the first step and that's owning value-producing hard assets. While the monetary value of the wind turbine and solar panels that power your home may increase or decrease with time, the value they provide to you doesn't. It remains constant. The comfort of a hot shower and convenience of food kept cool in a fridge and freezer will continue indefinitely. Hard assets make you more secure. Fifty acres of forested property, a building with a woodstove, a chainsaw and several jerry cans full of gas mean that you can stay warm all winter. A large rainwater catchment system and solar-powered pumps mean that you'll have water for your home and garden even during a drought. A large vegetable garden and tools to tend it, whether they are hand tools or a rototiller, will keep your family fed with fresh, healthy food at very little cost to you. Hard assets should form the backbone of your financial independence and sensible prepping.

Step Two - Barter

Step two in moving away from using dollars for exchange is barter. This is what humans have done for eons, but in recent history we have created an intermediary for this exchange called currency. A tradable currency, whether gold and silver coins or paper dollars, gave us more flexibility. We could barter with anyone, anywhere because we didn't necessarily have to be close to them. With the world reserves of oil depleting I believe we are going to see a return to a more local economy, and I think this is a good thing. You need to start thinking about what your current skills are and how you can improve them to barter.

Obviously if you are an auto mechanic you probably have a pretty good set of skills to use for barter. The oil isn't going to run out overnight and some people will still be driving cars. As the economic downturn continues people with cars will be hanging on to them longer and longer, and they'll need to be repaired more often. Some skills will be much easier to barter than others. If you have a strong back and a chainsaw you'll probably find people who have standing timber on their property but don't have the time or aren't in good enough shape to cut it themselves.

Chapter 19 discusses the importance of staying healthy, and bartering your services is one of the prime reasons. Our economy has increasingly moved to "information" jobs where we exercise our brains but not our backs. The economy will be returning to more physical jobs as we start to run out of the cheap oil that displaced all this manual labor. The skills that accountants and lawyers and information workers have will be much less in demand than doing real things that make it possible to stay warm and fill an empty stomach.

Step 3 – Non-traditional Tradable Goods

In the "hard landing" part of my workshops I take the demise of the paper dollar to the extreme and suggest that besides bartering your time and skills for goods you should have items that you can trade as well. What form this takes will depend a lot on where you live. One thing that is well known about economic downturns is that the consumption of alcohol increases in relation to the severity of the recession. I would suggest that since this is the worst economic upheaval since the Great Depression and it may in fact eventually be worse, alcohol is going to be a hot commodity. So it would be a good idea to have a fairly good stock of this. What type you have will depend on whose services you expect to need in the future. If you think you'll need lawyers and accountants you may want to make sure you have a well-stocked wine cellar. If you think you'll be needing repairs to your car and a good supply of firewood, you'll probably be more in need of whiskey. Beer is a good option as well, but it'll take up more space per unit of alcohol and may not last as well as wine or hard liquor, which might even increase in quality as it ages. If you're a drinker yourself, having a well-stocked wine cellar isn't a bad idea if you can afford it. Having a lovely meal of vegetables fresh from the garden with a nice glass of wine will make any bad economic reality seem perfectly tolerable.

While most of us know the health risks associated with alcohol and tobacco, many people still smoke and drink as a comfort during stressful times. A supply of these may be a good idea. It's a strategy adopted by many portfolio managers during recessions: they increase holdings in companies with products where consumption is in inverse proportion to the state of the economy. Governments tend to use the same strategy with taxation, often raising "sin" taxes on booze and cigarettes as a way of increasing revenue while decreasing consumption of legal items that are considered to have a detrimental impact on health. It's rare that sin taxes reduce consumption and tend to drive more activity in these products underground.

As times get tougher security becomes a bigger factor in people's lives. Also, some people who have hunted in the past will increase their consumption of wild game that in most cases is free. For both of these reasons ammunition, be it bullets or shotgun shells, will probably be a fairly valuable tradable property. I realize that for someone living in a suburban enclave much of this will seem quite foreign because they don't spend much time with people who hunt. But a percentage of the population does enjoy this activity, and others appreciate the low cost of the meat they get. If you've installed a wood stove in your home as a backup to natural gas you're going to need a source of firewood. If you don't have your own property to cut it from you're going to have to get it from someone who probably lives in a rural area where hunting may be the norm. So if paper dollars are starting to be less attractive as a means of exchange, being able to supplement your offer for the firewood with whiskey and shotgun shells may get you on the delivery route.

Ideally what you want to end up being able to barter is something that is renewable or that you can continue to produce over time. Food will always be in demand, so it's a good idea to look at producing the things you grow well in larger quantities. If you find that your soil grows abundant potatoes, then make sure you grow lots of extra. They store well and can form the basis of a sound diet. If you get the hang of keeping chickens, making the chicken coop twice the size you need isn't a lot of extra effort and eggs are a very popular commodity for trade. With chickens, once they're past their laying stage you have a source of meat to eat or trade as well.

Right now many of the items that we used to produce ourselves like clothes and blankets are very inexpensive. They are produced offshore in countries with very low wage rates, and cheap oil allows them to end up on our store shelves at very low cost. The end of cheap oil is going to mean a movement away from globalized trade and a return to more localized trade. So being able to turn cloth into clothing may once again be an option for trade. Quilts and blankets which in the past were often locally made may once again be something people will be willing to barter for. It is difficult to anticipate what your local needs will be, but over time they'll reveal themselves and you'll just have to have a keen eye to spot them. The day of the insanely cheap "Dollar Store" selling items for unbelievably low prices when so much work has gone into them may be coming to an end. The whole premise of this economic model is cheap and abundant energy, and those days are drawing to a close. We will once again be returning to a time when we need to depend on our neighbors for our economic well-being, and they'll need us. I believe most of us would rather know the people we trade with than employ some worker thousands of miles away while neighbors lack paying jobs. A fundamental change like this may be disconcerting but it's not a bad thing.

If you think this view of the future with reduced energy supply and a deteriorating economy is a possibility, I would suggest you start thinking about what you'll be able to trade in the future. The advantage of doing this now is that if you agree that hard liquor may be an excellent item to barter with, then now is a good time to start accumulating it. The price right now is probably much lower than it will be in the future, so it's a good time to stock up. Since you're going to be setting up a pantry for your food stores this will be an incentive to make it larger. Cases of liquor will store well and will be strong enough that you can pile other boxes on top of them which contain items you need more often.

Some of the items in your pantry may be very tradable as well, so consider a really big stockpile. Things like toothpaste and matches and soap will be very popular if there are ever disruptions to our supply lines. Toilet paper could become pretty valuable as well. You'll need a large space to store it, but if you've got a spare room somewhere, stocking up on something like toilet paper could turn out to be a pretty good gamble. It lasts forever, so if you can't trade it you can always use it yourself.

Step 4 - Precious Metals

Much of our current economic mess started when we decided to stop using the gold standard as a basis for our economies. Paper currency backed by a precious metal like gold or silver is the soundest method that humans have discovered for conducting economic activity. Any time you allow governments to introduce fiat currencies that aren't backed by something they will inevitably debase it and it will lose value. So if economies should be backed by gold and silver, is it a good idea for you to own it? Yes! It's a very good idea. In fact, it's one of the most important recommendations of this book. You should own gold and silver and you should own it now.

From 1933 until 1974 it was illegal for Americans to own gold. When Roosevelt took office there was a run on banks and people were withdrawing cash and gold. At that time dollar bills were still certificates which represented gold or silver that was on deposit to back the bill. Technically you could walk into a bank with your paper money and demand that amount in gold or silver. One of Roosevelt's first acts was to declare a "bank holiday" which closed all banks for four days. This was when he made the famous speech in which he said, "The only thing we have to fear is fear itself." In other words, if we all just calm down and stop panicking we'll be fine. When I reopen the banks we're all going to be mellow and we're all going to stop this crazy "run on the banks" and then we'll get through this.

Part of his Emergency Banking Act made it illegal for American citizens to physically own gold. The Federal Government confiscated all gold and if American citizens were caught with gold in their possession they faced a $10,000 fine, which was a huge sum in those days, and five years in prison. The Federal Reserve purchased gold for about $21/ounce and then informed foreign investors that U.S. dollars were backed by gold at the rate of $35/ounce. In 1971 Richard Nixon declared to the world that the U.S. dollar was no longer backed by gold, but he still wanted the dollar used as the "reserve currency" of the world, hoping central banks would keep U.S. dollars in their vaults rather than gold.

So for many decades it was illegal to own gold. But in 1974 the act was repealed and owning gold was once again legal. Owning gold has traditionally been viewed as a hedge against uncertain times or against inflation should a government choose to crank up the printing press and devalue the currency. In the early 1980s the price got up to around $700/ounce and it has fluctuated in the $300 to $400 range for many years. In March of 2008 as Wall Street started to implode the price hit $1,000/ounce and after a brief dip to the $800 range by March of 2009 it was back in the $1,000 range. Since then it has climbed as high as $1,900/ounce. It fell back down into the $1,200 range, which I think gives people a great opportunity to purchase before it goes back up.

This may seem outrageously expensive to you but many people argue that this is still actually extremely low. Based on the state of the world economy gold is probably very undervalued.

Gold is so rare a metal that all the gold ever mined could fit into a 20-cubic-yard block. Or as *National Geographic* magazine reported in its January 2009 edition "In all of history, only 161,000 tons of gold have been mined, barely enough to fill two Olympic-size swimming pools"[4]. It's a valuable metal for some industrial processes and has always been in demand for jewelry. It has always retained its value and been a safe haven for wealth invested in it during uncertain times. I hope Part I of this book helped to convince you that we are indeed in uncertain times. So you should be looking for a way to preserve your capital and wealth, and precious metals like gold and silver offer that. Most of us have heard casual reference to precious metals or commodities by financial advisers. They say it should be a small part of a "balanced portfolio" with the bulk of your investments in stocks and bonds. And how did that go for most people in 2009? Not very well, with many people's investment portfolios off by up to half.

So what I'm suggesting is that you take that small percentage that you've heard people recommending and make it a much larger part of your investments. I'm suggesting that precious metals offer the potential for great wealth preservation and for capital appreciation if you're looking to increase your wealth. First and foremost I must remind you that I am not a certified financial planner. I am just someone who for 25 years has tried to manage his money effectively. I started with mutual funds and then for a while I had some money in ETFs and some money in government bonds. Then in 2000 the books I was reading started to warn about the housing bubble which they called correctly. I would suggest to you that all those same people who years ago were screaming that we should get out of the market and that this crash wasn't going to be like any others we've experienced in recent times are now all screaming that it's time to own some precious metals. They're suggesting that with the economic mess we're in the likely outcome in terms of paper money is not a good one and it's time for you to start thinking about owning something of real value.

If you talk to your investment advisor about this he'll probably suggest you buy shares in companies that mine gold and silver. Remember, those shares trade on stock markets and if the market crashes they may go with it. If you go into your bank to discuss this someone may suggest you buy "gold certificates." This is a piece of paper that says the bank has put aside some gold for you and has it in safekeeping. Remember what I talked about earlier and what happened in the last depression. Gold was confiscated. The gold the bank has on deposit for you somewhere will be the easiest for a government to grab. So I would suggest you stay away from gold certificates. If you purchase gold coins or bullion the bank will suggest that you leave it in your safety deposit box in the bank. While this will probably keep it safe from burglars it will not keep it away from a government that wants to confiscate it.

If you buy gold and silver you should own it. You should have it and you keep it in your possession. There's no use having it and having the government confiscate it. What you're really doing when you invest in gold is putting your money in a place where you think it will hold its value. You should plan on keeping it for a long time. This isn't a short-term investment. This is a long-term move. This is the sort of acquisition that you are purchasing as a nest egg for an uncertain future. If you buy it at $1,000/ounce and it goes to $1,500/ounce you shouldn't sell it. If it goes up that much there is a reason, which is that people are losing confidence in the financial system. They are questioning the value of those paper dollars that the government is flooding the market with. So if it appreciates in value you should be even more determined to hang onto it. This isn't a "short-term make-a-quick-profit-and-get-out" strategy. This is a long-term hedge against uncertain times.

Since the economy started to unravel the demand for gold has risen steadily. Wealthy people have been moving more of their assets into precious metals. In fact the demand for gold and silver has been so high that it has often outstripped supply and is unavailable. Many sources including the U.S. Mint have found themselves without inventory of various precious metal coins since 2008.

How to Buy Precious Metals

Gold and silver comes in two main forms, bullion or coins. Coins are often referred to as "rounds." While coins may be more familiar to most of us I prefer to buy bullion in wafers or bars. A coin will often have a nominal "face value" such as the U.S. Eagle Gold Coin which has a face value of "Twenty Dollars" engraved on it. The Canadian Gold Maple Leaf Coin has a face value of "50 Dollars." This means the coins have a "legal tender" value of $50, but the numbers are largely symbolic because they don't represent the market value of the coin. So if you go into your local store to buy groceries the store is required to accept that coin as being worth $50. But that's ridiculous since it cost you $1,500 to purchase the coin. So the face value or legal tender value of a gold or silver coin is irrelevant and confusing. If in the future you want to use that coin to purchase something, will the person you're completing the transaction with use

the current price of gold, which could be thousands of dollars an ounce, or would they want to use the legal tender value of the coin?

The advantage of wafers or bars is that there can be no confusion about exactly what they are. In the picture you can see a 1 oz. gold wafer which is identified as 1 oz. of .9999 fine gold from the Royal Canadian Mint. There is no nominal face value. This is an ounce of gold and nothing else. If you want to sell your coins back to a bank in the future they will be valued according to the precious metal content. But if you are negotiating a financial transaction with your dentist there may be some debate about the value of the coin, so it is best to buy wafers if you can.

Coins are a bit easier to purchase and their disadvantages compared to wafers are minimal, so if the choice is coins or nothing, take the coins. You'll also have to decide if you want gold or silver. In my dreams I have rooms full of gold coins, just like Johnny Depp in the movie *Pirates of the Caribbean*. In real life I have never made a great deal of money and do not have much in the way of financial resources, so what precious metals I do have are silver. While an ounce of gold is worth about $1,500 today, silver is worth about $30 an ounce. Which means I can buy more silver. Gold is without question the premium and preferred precious metal to own, but with limited funds I've had to resort to the poor man's gold and very little of it. My choice has been to put money into solar panels and wind turbines because they make living off the grid much easier. I hope to purchase more precious metals in the future depending on my finances.

The other option I'll mention is older and collectible coins. Historic coins offer proof that coins containing precious metals maintain their value over time. The one type of gold that Americans could keep when Roosevelt was confiscating gold was historic coins because there was no way to properly value them. Also, since it was conceivable that some of the gold the government confiscated might be melted down to produce larger gold bars, the use of coins was viewed as destroying historical artifacts. From that perspective these coins do offer some benefits. The challenge is that buying them is a very specialized skill and not something the average person should undertake. If you are interested in this you need to research it thoroughly and read everything you can on the subject. Then you'll need to find a dealer you trust to help you as you pursue this option.

So your choice is based on how much money you have and which you prefer. Both gold and silver have actual industrial uses. Silver is used in electronics as a catalyst in chemical reactions, in mirrors, and in silverware for the table. At one time it was used in dental fillings and in photographic film. Gold can also be used in electronics and jewelry, so which of the two will be more in demand for industrial purchases will change over time. The benefit of silver for many of us is that it is less expensive and we can buy more of it for the same amount of money. The relative price difference between gold and silver has remained fairly consistent over time.

The other benefit of silver is its ease of use in financial transactions. If in the future the value of paper dollars becomes questionable and you use your gold and silver for transactions, the lower value of silver will make it easier to use from day to day. If you need some dental work done, a dentist accustomed to charging $1,500 for the procedure might be comfortable with a one-ounce gold coin. If on the other hand you want to purchase 20 pounds of flour from the local farmer which she values at $100, a $1,500 gold coin is going to be difficult to use. Do you try and cut one-tenth off the coin to give to her? A silver coin worth $25 or $30 is going to be much easier to use. So having a mix of gold and silver if you can afford it is a good idea. If you're determined to own gold on a limited budget or want smaller amounts to better facilitate transactions there are 1/10 ounce gold coins available now. I just went to the U.S. Government Mint website to check on the 2012 American Eagle One-Tenth Ounce Gold Coin and it was "Sold Out." www.usmint.gov. They sold out of silver coins a few months ago. Hmmmm.

Where to Buy Precious Metals

There are lots of places to purchase gold and silver, from banks to coin dealers to websites. One thing I recommend is that if you are buying precious metals as a long-term store of wealth and potentially a hedge against very challenging economic times, try to purchase with the minimal number of audit trails. Essentially try and purchase precious metals for cash with no receipts. You walk into the store or coin dealer, pay cash (dollar bills not plastic) and you walk out with the gold and silver. I suggest this simply from a capital preservation standpoint. If the government decides it wants to get the gold back, it's better if it doesn't know you have it. If you purchase it from a bank that fills out an order form, or online with a credit card, there is going to be a record of that transaction. There is absolutely nothing illegal about taking dollar bills to a coin dealer and purchasing precious metals. I suggest this is the way you build your precious metal reserve.

If you can find a local coin dealer who will work this way that's excellent. Get to know him, but on a first-name-only basis. He may be able to guide you through the process and give you additional insights into the various products available. He may also have access to private sources for these items rather than having to purchase them directly from the Mint. He may let you know when he's purchasing a collection from someone so you can take a few days to get some cash together. Building a collection of gold and silver should take time because it's not always the best idea to walk around with a huge amount of cash.

Try the yellow pages under coin dealers. You can also buy your coins directly from the United States Mint at www.usmint.gov or the Royal Canadian Mint at www.mint.ca. Obviously there will be a very noticeable audit trail to your purchases there. The U.S. Mint website does have a feature that allows you to find local dealers in your area (www.usmint.gov/bullionretailer). There are also independent online dealers like Border Gold (www.bordergold.com), Kitco (www.kitco.com), USA Gold (www.usagold.com), and Goldline (www.goldline.com). Companies change over time and offer different terms and buying strategies, so do some research to find a source you are comfortable with.

How to Store Precious Metals

The first thing people think of with items like this is that they should put them in a safety deposit box. That traditionally has been sound advice but may not be appropriate for times like these. The question is whether, with the severity of the economic downturn and the Patriot Act's new powers of personal intrusion, it makes sense to have an asset you've purchased for a worst-case scenario in a place where you may not be able to get to it. I would suggest this is not a sound place for your gold coins or at least all your precious metals. I would suggest you store them in your home.

A safe would be a good place to start. Some would suggest you'll need two safes. The first, where you keep a bit of cash, is the one you take the home invaders to. The second, better-hidden safe is where you keep the bulk of your assets. Not owning much silver it's not much of an issue for me, but I prefer to use the camouflage technique. On average burglars will spend eight minutes in a home. They're going to go to all the easy places like the drawer in the table nearest the door where the man stores his wallet, then upstairs to the top drawers of the dressers, then to the jewelry boxes on top of the dressers, then the night tables, that sort of routine. What you need to do is disguise what you don't want them to find. Still have a big collection of VHS movies? How about taking that

Gone with the Wind tape—well not that one, that's a good movie. How about that movie you always regret buying because it was a train wreck? Put your silver in there. If a burglar is zipping around the house looking for valuables, VHS movies don't exactly scream "I'm valuable," and if you have a wall full of them who's going to bother to take the time to check?

There are lots of commercial products you can buy for this camouflage strategy: pop cans, flower pots, false electrical outlets. You may want to check them out and see if anything twigs. While it might be tough to make your own mini pop-can safe, next time you open a canister of potato chips you might want to peel the foil protector on the top back carefully so that once you've eaten the chips you can put items in the container and put it on a crowded shelf.

In my workshops I use a photo entitled "Spot the Silver Bars"; it shows a messy, chaotic paint shelf in the garage, which is a bit of disaster. Humor is important when discussing issues like this and my lack of garage tidiness skills always get a laugh. I kid that there's someone in the back row text messaging their cousin to get over to my house because I've got silver stored in old paint cans. I come clean that I don't have much silver and what I do have is not stored there. The example is simply to show the value of camouflage.

A recent magazine article traced a couple's attempts to bury valuables in their backyard. I know, this sounds crazy, but I'm throwing it out there anyway. Humans have been doing this for centuries so apparently there is some merit to it. If you live in an urban area you need to do this discreetly, preferably very late at night. If you saw the Alfred Hitchcock movie *Rear Window* you'll know it's important to make sure some Jimmy Stewart-like neighbor isn't observing your late-night activities in the garden. If you live in a cold climate remember that once you get a good frost in the ground it will be difficult to dig your stash up, so anything you bury should be for the long term. If you live in Minnesota and decide it's finally time for a big-screen TV just before the Super Bowl in January then a backyard wealth deposit is probably the perfect antidote, because without a large propane torch to heat through the frost that stash is staying buried until the spring. This is an excellent method to prevent you from being tempted to use the stash unless it's a real emergency. I extend my apologies to my male readers who would argue that a new widescreen TV for the Superbowl does in fact constitute an emergency.

I would not suggest you bury cash. It is next to impossible to keep it dry and it will deteriorate over time. Bury the gold and silver. Since moisture is the enemy having the container as airtight as possible is critical. The photograph shows an easy-to-make and relatively inexpensive container. You take a length of IPEX 75mm rigid plastic plumbing pipe and use two of the screw-on caps that a plumber would use for a cleanout. A plumber would use only one. You use Y2 ABS Solvent Cement to glue the cleanouts on each end. Once they've dried you can put some Teflon tape on the thread of the screw-on cap to help make it extremely airtight. Use a large wrench to tighten the cap firmly once you've placed your valuable inside. Then it's time to head out to the yard after *The Late Late Show* is over and quietly dig a nice deep hole to put it in.

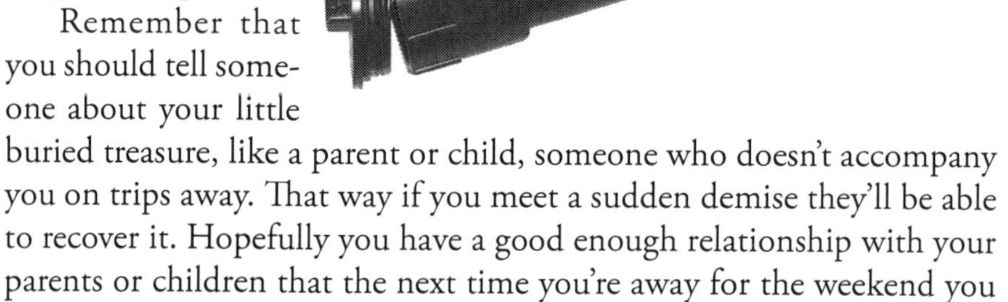

Remember that you should tell someone about your little buried treasure, like a parent or child, someone who doesn't accompany you on trips away. That way if you meet a sudden demise they'll be able to recover it. Hopefully you have a good enough relationship with your parents or children that the next time you're away for the weekend you won't come back to find that someone's been digging in your backyard. You need to evaluate your own family dynamic critically before you make this decision.

I understand that many readers right now are shaking their heads and saying "this guy is wacko." I understand and accept your criticism. I seem very outside the mainstream. Or at least I seem very outside the mainstream today. With an historic look at things though, my approach is in keeping with the way humans have dealt with trying to store wealth over the centuries. All this talk of gold probably seems very strange to you. You're conditioned to hearing that the people buying gold are the same ones who are building bunkers in their backyards and stocking up on shotgun shells and camouflage gear. I know it seems strange to someone raised using paper dollars. It must seem even stranger to someone who's quite young and has hardly ever used cash at all and has relied on electronic transactions with credit and debit cards. It is very much in keeping with my theme, though, of returning to the values of your grandparents and the independent way they lived. In the 1920s and early 1930s they could go to the bank and withdraw gold for transactions. It wasn't necessarily convenient, but it was an option. Like my recommendation that you return to a time of energy and food independence, having gold and silver coins around is just another step on that road to personal independence.

I would suggest that precious metals have historically been a sound method of wealth preservation. It has also been an excellent means of self-preservation for many in historically dangerous times. As the Nazis rose to power in Germany in the 1930s it became clear to some Jews that this was not going to be a good place to stay over the long term. Many emigrated early on while it was still possible to do so legally. Those who waited until later in the decade found it increasingly difficult to leave as anti-Semitism rose to a fever pitch and the movement of Jews was restricted. Those who did get out were able to bribe their way out, and with the German currency having been severely inflated during the decade the best way to bribe someone was with something that maintained its value—gold and silver. Gold and silver coins and jewelry were what allowed some to escape. I certainly hope it will never come to that here, but since owning some precious metals is part of a sound financial portfolio, it's nice to know that if you have them in your possession they can save you in an emergency. This is sensible prepping.

Once you get over that "survivalist" mentality about precious metals and realize they are a part of any household's financial assets, they'll suddenly seem less strange and more comforting.

I'll add just one final note on this topic and suggest that you may want to keep a bit more cash on hand than you do now. I realize I've been arguing that inflation erodes the value of paper money, but it still is the main medium of exchange in our economy. Debit and credit cards have allowed many of us to function perfectly well in the economy without carrying any cash. This will still be the case as long as the power stays on. Earlier on I spoke of the challenges facing our power grids and how the lack of investment in the electrical infrastructure has increased the likelihood of power disruptions. While you're going to be putting up solar panels with a battery backup to power your home, you may still want or need to purchase things during a blackout. If the power goes down and you're out and about and have no cash in your wallet you are not going to be able to purchase anything. I recommend that you always keep your gas tank at least half full, and anyone caught in the traffic nightmare of trying to evacuate coastal areas during hurricane watches knows you should be starting with a full tank and filling it when it hits halfway. But if you were a distance from home and your gas was low and the power went out, if you could find a gas station with a backup generator to pump gas it sure would be nice to be able to buy it. Cash—dollar bills—might help.

Some would recommend that you have several months' worth of cash on hand. Once you've calculated your budget you can decide on an amount. I'm not really contradicting what I've said about precious metals. If there is a "bank holiday" declared, people will still be completing transactions with cash. There will be an interval before people realize that dollars have lost their value and they switch to gold or silver. So if there is a period of dislocation, having lots of cash will help. If you need prescription medications or your child needs an asthma inhaler you can use that cash reserve to stock up. If things in the economy are unsettling to retailers they will be much less enthused about electronic payment and much happier to have cash. That moves you to the front of the line.

If you needed to stay in town overnight because the ice storm that caused the blackout made driving treacherous, it would be nice to be able to pay for a room in a hotel or motel. Yes, these are unlikely scenarios but they could happen and not having $50 or $100 cash in your wallet could turn an inconvenient situation into a real mess. Keep some cash in your wallet. Keep some cash at home. I don't believe governments will have to declare "bank holidays" and restrict your access to your money, but if they do you'll appreciate having some cash on hand. When the FDIC takes over a bank you don't want to be one of those stressed people standing in line in the heat as they did at IndyMac trying to get out some cash. Take it out now. Put it in your pop-can hiding place and leave it there. Don't use it for the pizza man when you don't have cash. You shouldn't have ordered the pizza if you didn't have the money at hand. In fact, you really should be making your own and taking the $20 you saved and putting it towards your mortgage. But you've heard all that by now and you have a new commitment to get debt free, which is the key to your new independence!

(Endnotes)

1 *William Bonner, Empire of Debt, John Wiley & Sons, 2006, page 328*

2 *Ron Paul, "What the Price of Gold Is Telling Us," LewRockwell.com, April 17, 2006, http://www.lewrockwell.com/paul/paul319.html.*

3 *Kevin Phillips, "Why the Economy is Worse Than We Know," Harper's Magazine, 1 May 2008, http://www.mindfully.org/Reform/2008/Pollyanna-Creep-Economy1may08.htm.*

4 *National Geographic, The Price of Gold, January 2009. Pg 43*

23 The Apartment Prepper

I'm writing this chapter for all of those people who live in apartments. Most prepping materials tend to assume you own a home and can put a concrete bunker in your backyard. And yet a huge portion of the population lives in apartments. And this is a very good way to live. You have a small area to heat and therefore have a smaller impact on the planet. Fewer of your walls are exposed to the elements so there is much less heat loss through them. Many apartment dwellers live in urban areas with easy access to transit, so they tend to not drive as much, which is also good for the planet. So if you live in an apartment, good for you! Keep up the great work! Mother Earth thanks you!

Dealing with a major upheaval though is going to be somewhat more of a challenge simply because you probably don't have as much space to store stuff.

So step one in your apartment prepping should be to "declutter." There are lots of shows on TV that explain how to do this, but ultimately you just need to invest the time. And set priorities. That box of high school essays might represent a huge investment in time but it's probably time to let it go. Or store it at your parents' house.

After your decluttering you're probably going to have closets full of empty space to start accumulating your prepping supplies! .... NOT! I know, it doesn't matter how much you try, apartments were never created for preppers.

So you're going to have to set priorities. And let's pick a relevant period of time you want to be able to "shelter in place." We've already experienced widespread blackouts that effect major metropolitan areas, but luckily they generally haven't lasted that long. Some areas of Manhattan that had previously experienced the 24-hour black-out in 2003 were hit by Hurricane Sandy and found that a few days into a prolonged blackout was a much different ball game. And it wasn't the dead of winter yet.

So why not make provision for about a week without utilities. First and foremost you should think about keeping warm. If you're used to sleeping in one of those apartments that are kept at tropical temperatures then you may not be prepared for a week without heat in December. So get a really good sleeping bag. Staying warm is one of the keys to dealing with an extended powerdown.

The next priority to consider is water. When Hurricane Sandy was approaching there was lots of warning. They knew days in advance that New York was a possible target. That afternoon the trajectory was clear. So, there would have been time to fill your bathtub up with water. I would have cleaned it, and then filled it up, so this water would be "potable" or drinkable water. If you have two sinks in the kitchen, fill one up with water. Hopefully you put aside a few buckets ready for this, so fill them up in the bathroom ready to use to flush the toilet. In many stores you can buy "accordion" type water containers that fold up very small for storage. If you have some of these in your emergency kit then you could be filling them up as the storm approaches.

It would be good to have several cases of bottled water, or gallon jugs filled with drinking water. Or more than several cases of water. How about 5? How about 10? You live on the 4th floor and there's no elevator? What a great opportunity to get in shape. Who needs a health club membership? Bringing home a case of water on your wheeled cart, then hauling it up those four flights of stairs should qualify you for the Marines!

Water is critical so you need to have a week's worth put away.

You'll need the same emergency kit that I discussed in Chapter 6. I would have lots of hand sanitizer in it since you're going to assume you won't have water to wash your hands with. There is a good chance that your city's water utility may have enough backup power to keep water flowing to your apartment, but you can't count on it. And since storms can easily put sea and contaminated water where you don't want it, you may not be able to trust the water for drinking. And if there's a "boil water advisory" in place but you don't have any way to heat it, then it's not much good to you.

So that one prepping closet you have is now filled with water, your emergency kit and that darn sleeping bag you've never used. Yet.

So that just leaves food. I'm constantly amazed at how little food many apartment dwellers keep on hand. I'm sure this comes from constantly being around food in an urban environment. You are always walking past stores with food, restaurants, food carts ... it's just always there so it follows that you'd be lulled into assuming it will just always be there, in abundance. It seems to be one of the things that the media likes to focus on during an event, how quickly restaurants run out of food or can't prepare it, and how much stored and refrigerated food quickly goes to waste. So yes, your strategy during a black out could be to watch for local restaurants that have outdoor grills cooking up food and giving it away before it goes bad, but I wouldn't suggest this be your preliminary plan.

Your first focus should be to actually have a week's worth of food in your house. And if you want to get a little crazy, it's not too hard to aim for a few months' worth. In a city you may prepare your meals fresh every day with fresh ingredients, but lower your standards a bit and start putting aside some canned goods. The kind you can just open and eat. Remember your standards are likely to be greatly reduced in these circumstances. And yes, canned beans or canned pasta will not win any culinary awards, but they will fill up your tummy and give your body sustenance. There are people who live on canned spaghetti!

Stocking up on canned goods is a great backup plan. If an extended blackout hits, start with the fresh stuff in your refrigerator. Open the door to the fridge and freezer as little as possible. Go in with a plan and close it as quickly as possible. Stuff in the fridge will start to lose its appeal within 24 hours and even if you've been able to keep the freezer door closed items will begin to thaw within 48 hours. You might not have a way of cooking what's in your freezer anyway. If it's meat and you have any doubts, don't risk it. If it's perogies that have thawed you'll probably be all right.

Next you'll have bread to eat up. If it's from a bakery nearby or whole grain it won't last as long as those loaves of white bread you have. So good news kids, another night of BP&J (peanut butter and jelly) sandwiches for dinner again!

Then you start moving to your canned goods. You'll notice I haven't discussed 'cooking' meals because that's really the Achilles heel of the whole apartment food-prepping question. If you have a gas stove, the gas utility will probably be able to keep gas coming to your apartment. This is awesome! Not only can you cook your food, you will get some heat from the stove and you can boil water. Your emergency kit must have matches. Lots of them. And since the electronic ignition will not work on the stove, you'll just need to turn the gas on to the burner and light it carefully with a match. For safety reasons you won't be able to get the oven on this way, so you won't be baking any cakes during the blackout.

If on the other hand your apartment has an electric stove, you won't be able to cook on the stovetop. If you have a balcony with a bar-b-queue you're lucky, but remember, you cannot "grill" inside an apartment or home, whether it's gas, propane or charcoal. It will produce potentially lethal carbon monoxide, which could kill you. Some people might suggest that you could use a smaller propane or kerosene camp stove but I would still recommend against it. It is often difficult to get the proper ventilation in an apartment so once again, it could kill you. Plus these stoves often don't have proper shutoffs so they will keep working if knocked over and could vent gas into the apartment if the flame is blown out, which would be a potential source of an explosion or fire.

Instead I think something like a "Sterno Stove" would be an excellent idea. www.sternocandlelamp.com. Sterno is made from denatured alcohol, water and gel so is safer to use indoors. You may have seen these small cans of fuel under warming or chaffing tables at buffets. They are safer simply because they produce a nice consistent low-level heat. A stove like this will be excellent for heating foods like canned soup, and what would be nicer than soup during a crisis? You can use it indoors safely and an 8 oz. can of Sterno will last about 2 ½ hours. So if you had 4 or 5 cans of Sterno you should be fine for a week's worth of soup warming or water warming. No, you won't be cooking any Thanksgiving turkeys on a sterno stove, but that's not the goal here. Sterno stoves are available in many hardware and department stores. The kit in the photo earlier in the book comes with some good quality candles for light, so it's a valuable emergency kit to have.

As I recommended in the heating chapter you should make sure you have working carbon monoxide detectors in your apartment. A natural gas stove could malfunction and lead to dangerous levels of carbon monoxide. And I've brought this up because people do crazy and dangerous things during a crisis, like using kerosene and propane heaters inside to get warmed up. If you did use one of these, which you should NOT, you have to ensure proper ventilation in an apartment. But you can see the problem with this; you're trying to keep warm by using a heater, but to use it safely you need to have the window open. So don't even try. Make sure any windows or patio doors to your apartment are closed tightly. Then put on your thermal underwear, layer on some sweaters and suck it up. Better to be cold than dead. While she probably didn't realize it, that old kerosene heater that your Aunt Gertrude gave to you could be lethal.

This final point, the reality of cooking and staying warm in the winter in an apartment, brings us to our final discussion on apartment prepping, and that's when to bug out.

This is going to be a personal decision and one with multiple inputs to consider. The information you use to make this decision is going to be based on your prepping. Hopefully you have a wind up radio and it has been essential for the last few days without power. It's allowed you to have realistic expectations about the power. After Hurricane Sandy, Conn Edison told people that they were going to be without electricity for days. It's important for a utility company to be honest because if you keep thinking the power will come back on at any minute and it stays off for days, rather than just accepting the situation, your mental state will be worse than if you got the facts.

Your reasons for leaving will be unique. It could be that you just can't stand another day without the internet or a working cell phone. You will never understand how addicted you are to that instantaneous information flow that modern technology provides, until you are denied it for an extended period of time. Some people will go into withdrawal within hours. For others it could be the isolation. On the first day there may be lots of activity on the streets as people survey damage and get used to their new reality. But this often ends once the sun goes down and those same streets are dark and unfriendly without that artificial daylight that electricity provides.

You may have friends or family not too far away that still have electricity and heat - in other words, civilization! So that bug out bag you prepared in Chapter 7 is now going to come in handy. You have to assume you may have to wait a while for transport out of the city, be it bus or train. You should also assume there may be a lack of consumables in the bus station, so take as much individually packaged foods like granola bars, and water, as you can carry. If you have a car hopefully you've been keeping it as close to full as you can so that you don't have to line up for gas, if there is any accessible to you.

Assume your trip will take longer than normal. Wear your cleanest clothes and wear lots of layers. Cold is your enemy, so dress for the worst-case scenario. Then you can peel off layers if you get too warm and put them back on if you get chilled later. Wherever you hope to catch the bus or train from may not have power, may not have heat, and you may be there for an extended period with other people who have decided it's time to leave too. So plan on staying warm. Extra clothing can also make a great pillow if you're stuck waiting somewhere.

I feel like I should talk about security here, but I'm not sure I'm comfortable doing so. If you're in Nashville and have legally acquired a carry permit, you can take your handgun with you. If you're in New York City, which has much tougher gun laws, it is unlikely you'll have a carry license, so you'd have to evaluate the wisdom of taking one. If there was a strong police presence as they try and maintain law and order, and they discovered you had an unlicensed firearm, they might not take kindly to that.

Then I think, well, there are some amazing handgun replicas, which are very tough to tell from a real gun. So you could take one of these and have it visible if you encountered a situation where you wanted others to know you were armed. But again, law enforcement people are not big fans of these replicas because they can't tell them apart either, so as far as they're concerned, you are armed.

You can always tell someone who isn't a handgun owner. If I was I'd probably be much more comfortable with the no-brainer of suggesting you take your handgun. But I don't own a handgun. Where I live you need a permit to own one, and you can only own one if you use it for target shooting or are a collector. You need a special permit to transport it to and from your shooting range. And yet criminals in big cities near me seem to have lots of handguns and are shooting each other and innocent bystanders with them all the time.

So there you have it, my wishy-washy stance on taking a firearm with you when you bug out. If you're bugging out after a number of days of chaos, then it's probably a good idea to "take 'em if you've got 'em." On the other hand, after several days there is a good chance the National Guard or other military units may have arrived to keep the peace, and things may be quite safe. You'll have to get a read on what's happening and govern your behavior accordingly.

There's no reason that finding yourself without power after a storm in an apartment has to be a catastrophe. Your apartment is going to stay warm much longer than a house with 4 exposed walls. With some simple preparations you should be able to ride out a few days of disruption. Focus on food, water and staying warm. Stay informed about what's happening. Have a plan. Have a good bug out bag ready in case you decide to leave. A couple of days of thought and preparation will allow you to mellow out about the possibility of dislocation in your city. Now you can relax. Head out for a coffee. And a Danish. And put a good, small LED flashlight in your purse or jacket pocket. Now you're a prepper. A 'sensible prepper.'

24 The Power of Community

I've spoken about 'community' in a number of other chapters including the chapter on security, but I decided to devote an entire chapter to this theme because it is so important. I wanted to make sure that if you only read the Table of Contents you'd recognize the importance of community as one of the keys to resilience and emergency preparedness.

I also think it's one of the biggest challenges for many people today. The way we live today often isolates us from neighbors. People come and go from their homes or apartments and don't take the time to get to know their neighbors. Often we tend to fear them, and figure it's better to just do our own thing rather than engage them. "There was that time I chatted with a neighbor but she started complaining and telling me her life story and I couldn't get away from her…"

Technology also has a huge impact on how we interact with others. Many of us probably know people in far off parts of the country or the world thanks to the internet, much better than we know the people who live in the house next door. And yes, you may have more in common with that fellow quilter or hockey fan in Europe, but in an emergency that could quickly change and you'll be in the same boat as your neighbor.

So it's a good idea to at least know your neighbor by name and be friendly to them. You don't have to become life long friends, just acquainted enough so that if you're heading down the stairs in your pajamas when the fire alarm is going off, you don't feel that uncomfortable. Well, you might not be comfortable in your pajamas in front of anyone, but you know what I mean. Knowing the people who live close to you could be very beneficial in the future.

I live near a small village where a number of groups of people are very active in trying to make it a great place to live. There is a group that organizes the Santa Claus Parade and a craft/gift sale in the library on the same day. The Lions Club, among other projects, hosts a breakfast with Santa the day after. There is an Economic Development Committee that has undertaken a number of improvements in our community. We have a large celebration on "Canada Day" (our 4th of July), which includes a parade and festivities at the ball diamond and fireworks after dark. We have several church groups. There is no shortage of groups to get involved with and they all rely on volunteer help.

It's no difference in cities, in fact, there are probably even more options and more likely you'll find a group of people that you share interests with. I started playing "Happy Hockey" in town during the winter and I've met a whole new group of neighbors.

For me there's something special about living in a small town and going to town and taking twice as long as you intended to because you meet someone in the grocery store, or hardware store, or on the sidewalk in front of the Post Office and you gabbed. I make a point of saying Hi to everyone I pass on the sidewalk when I'm in town, and the more people I can say "Hi Bob" to, the better.

Even when we lived in a large city we made a point of getting to know our immediate neighbors. Our street had about a dozen houses along each side of it and we made sure to know the names of each family and find out something about them. We went door to door to organize events, like street-long garage sales and once we even invited all 24 families to our house for a potluck dinner on the front lawn.

It takes a long time to become part of a community. And you have to work at it. But I think it's really important. There is no way to replace the importance of being part of a community in uncertain times, but it takes time to become part of one. And it requires effort. And there is never a shortage of groups in any town, small or large, that needs volunteers and people to become active in them.

The groups that I've been active in have been focused on improving the community, through economic development, green projects, parades, you name it. My community is a great village to live in because of the people. When a house burns down or there is a challenge with a family in our community, these groups step up and fundraise and make sure that the person or family in distress feels the village standing behind them.

If I were living in a city right now and feeling a bit antsy about rising oceans or European bank holidays, I'd be finding a group to join. I'd be learning to quilt, or play softball, or I'd find a garden plot, or a "Transition Group" and I'd force myself to start going to meetings. And I'd take on a task when the call for volunteers goes out. And the next thing you know I'll be a coffee shop with a bunch of new friends figuring out how to save the world. Wow, that was easy.

25 The 'Happy' Sensible Prepper

Think back to a time in your life when you were really happy. Maybe it was playing with some friends when you were just a kid, out on your bikes or at the park. Maybe it was swimming in a pond or a lake on a hot day. Your first long walk with someone you really liked. For many of us, great memories aren't necessarily those that involve money. They involve non-monetary things: friends, places, experiences, just being somewhere.

Now if all you can come up with is events related to money—your first house, your first car, your first paycheck—you've got to try and think of some other things you've enjoyed that didn't involve money. This is really important because you have to try and break that very strong link in your mind between money and happiness.

If you were fortunate enough to grow up in a stable environment as a child, think back to some time that you were enjoying yourself. This is hard, because we're all prewired to much more easily remember negative experiences. Getting scolded by your parents, a bad mark on an assignment at school, being picked on by a bully. It's only natural and in some instances a survival instinct for these memories to be stronger than happy ones. When an ancient human was chased by a saber-toothed tiger it was important to remember to take evasive action the next time she saw one in the distance.

I think if you try really hard you're going to remember some good times that involved little or no money. I know that when we got married in the early 1980s, at a time of relatively cheap gasoline, Michelle and I drove out west for the summer for our honeymoon. I had dropped out of university and Michelle hadn't found a full-time job, so we got married in June, jumped in our tiny Toyota Tercel, which got insanely great gas mileage (and wouldn't pass any crash test today), and drove from Ontario across Canada to British Columbia and then down to California.

We stayed with family in Edmonton and Vancouver. We camped most nights in provincial parks, which were very cheap in those days before the neo-conservative age of cost cutting reduced money to public parks. The odd time after a number of days of rain we might find a very cheap hotel to dry out in. We cooked our own meals, stayed out of tourist places, and even camped for awhile on a free beach in California. We probably spent around $1,000 the whole summer and it was fantastic. Every day was a new adventure—driving somewhere we'd never been, seeing things we'd never seen—and we hardly had a penny to our names.

We certainly had no security. We had no jobs to come back to. We had very few possessions and most of our wedding gifts had been coolers and camping equipment for the trip. But regardless of how little money we had, we were happy. Every day was an adventure and our biggest challenge was figuring out the cheapest way to eat that day. Thirty years after that trip I can still remember many of the experiences as if they just happened. The sounds, the smells, the feelings. Lying in the tent with the flap open first thing in the morning. The anticipation of what was around the next corner.

Have you had an experience like this? Maybe during the last few years you've been living in a large house with lots of rooms and cars and the expenses that go along with them. Yet when you think back to that first house you bought, or that first apartment you rented, the one where you used old wire spools for coffee tables and made bookshelves out of discarded wood and bricks, you probably have some pretty good memories.

Now you have to reach back to those times to remind yourself that your happiness isn't necessarily a product of your income. It's not related to how big your house is or what kind of car you drive. If your work and living and driving arrangements have changed or are in transition, perhaps to a more scaled-back level, this is not a bad thing. In fact, it could be a good thing. It could be a really great thing if you think about it in the right light. It's up to you. You can sit around feeling sorry for yourself, missing that 4-bedroom, 3-bathroom, 3,000 square-foot monster home in suburbia, or you can start looking at the upside of your new living arrangements. Smaller places require much less cleaning. They're easier to heat and cool. You don't have to work so hard to pay for them. If you're back renting as the housing market declines, you'll be able to buy in with much saner prices in the future. This is a good thing.

Buddhism talks about the concept of "mindfulness" or being aware of one's thoughts, actions, and motivations. For people suffering from depression mindfulness can be used as a strategy to help them try and remove themselves from the sources of their stress and become very focused on the here and now. For many the stress or depression comes from bad experiences in the past or fear of future unknowns. By separating yourself from those things that do not impact you at this very moment you can try to start bringing some positive light back into your life.

You have to start "living in the moment". This just means you have to tell yourself that what you're doing right now—having a coffee with a friend, weeding your vegetable garden, loading firewood for next winter, cutting vegetables from your garden to make soup— these are good things. Right now, I'm happy. I will not dwell on what might have been if the economy had continued as it had for so many years before I lost my job. I will not worry about what's going to happen months from now. I can't change the past and I only have some control of the future. But I can tell myself that what I'm doing right now is a great thing. I'm taking joy from this everyday activity.

We live in a society where you need money for just about everything. The goal of this book has been to help you start living more and more as though you're going on a canoe trip and you have to pack everything you need for a week in your canoe. Where you lock your wallet away for extended periods of time because you don't need it, because you grow your own food, make your own electricity, and generate your own transportation by converting your food energy into motion on your bike. The more elements of your life you can free from the money requirement, the less burdened you'll be and the less stress money will cause you.

Studies completed since the 1950s prove there is no link between income and happiness. Since 1950, as income has exploded and material wealth has gone up exponentially, the number of people who consider themselves "very happy" has gone down steadily. We now have almost four times the material wealth we had in 1950, bigger houses, more income, and more multiple storage units for our "stuff," but we aren't any happier. Ask a person who has traveled the world who seems happier, North Americans or people with significantly less than we have.

I have found that being grateful for what I have has always made a huge difference on my outlook. From a big breakfast to a cup of tea with a friend I believe the more emphasis you can put on appreciating the small things in life, the more likely you are to find yourself contented and happy. And I've found as I work more on Maslow's lower needs like growing food and staying warm by cutting my own firewood, the more

I've realized those higher needs like 'self-actualization.'

This one-time gift of fossil-fuel energy that we have been burning through in the last 100 years is starting to show signs of dwindling, which means that we are going to start spending more of our income on basics. In fact, as 7 billion people compete for less and less food, I believe that shortages are a possibility and that we'll all be even more focused on food. As we start to have price spikes or shortages of the fuels that power our lives we're also going to become more focused on shelter and will be devoting an increasing amount of our incomes to that. So it looks as though much of the attention which we previously focused on needs higher on the hierarchy is going to be focused on the lower, more basic needs.

"Prepping" can seem like a sort of dark thing to do. It's like you're preparing for the end of the world as we know it. And yet all you're really doing is putting food and things aside for a rainy day. It's the way humans have lived since they settled down and started farming. Ancient ruins always include grain storage rooms. Our ancestors stored some to eat and some for seed to plant the next year. So all you're doing is carrying on a tradition of human civilization. What we seem to miss is that modern society has so cushioned us from the ebbs and flows of nature and the elements that we don't have to really think much about the future. Someone else will provide us with the fuel to keep our home warm, to keep the lights on, and keep food on the table. Someone else is storing some grain for us, right?

Living off the electricity grid, Michelle and I have had our share of dislocation over the years. Flooded buildings, a phone system out of commission for extended periods of time, a precipitous drop in our income with the economic collapse in 2008. It's not always fun. It's stressful. It throws you for a loop.

What has eased our stress levels has always been our rainy day preparation. Our root cellar full of potatoes and onions, our pantry full of pastas and rice, some savings, next years' firewood stacked and ready to keep us warm. It's also helped that we don't have any utility bills. No natural gas or oil bill for heating our home. No bill from the electric company. And if we could just give up coffee (and the occasional Oreo) we'd hardly ever have to spend money at the grocery store. Sure our diet would be bland, but we'd be in control of it.

There are lots of examples of people who had a plan for surviving very challenging situations. I've read about people who survived horrific plane crashes. Often they say, "I sat down at my seat and checked out the emergency exits. I found the closest exit. I said to myself, "If this plane goes down I'm getting to that door and I'm getting out that door. Nothing will stop me." And when there was chaos and smoke they got to that door and they got out. Many people who survived the WTC attacks knew where the staircases were because they did lots of fire drills. It turns out that fire drills save lives. Some people knew the staircases in those towers so well that when they got down from the floors above where the planes had hit and the staircase was blocked they went back up and came down another set. They had made a plan and they stuck to it and nothing stopped them from following through.

Make a plan. Practice your plan. Rehearse your plan. Stick to your plan. You can be flexible with your plan. Things will change, but once you set that goal stick to it. Whether it's getting out of that burning plane, an approaching windstorm or getting out of debt and buying a hobby farm in the next five years, set your sights and stay on course.

Next winter my lights will be on, my fridge will be keeping food cool, my freezer will be keeping vegetables from my garden frozen, and my computers and Internet will keep me linked to the outside world 24 hours a day, 7 days a week without the possibility of any interruption, because all the power is generated by my solar panels and wind turbine. Next winter I'll be having showers and baths with water heated by the sun and my wood stove, regardless of whether the propane truck decides to come all the way down my road to fill up my tanks. I get to stay clean without being at someone else's mercy

The technologies exist for you to be very independent. You probably have enough income to create an emergency kit for a natural disaster. There's also probably a pretty good chance you have enough income to start taking a look at the big picture and investing in technologies that make you more independent. They don't have to be high tech. A wood stove is pretty low tech, but new models are amazingly efficient.

A 'sensible prepper' has a plan for an emergency. They are also in the process of making themselves more resilient to some of the bigger challenges coming our way. They're just putting some grain aside to plant next year like Egyptians on the Nile did 10,000 years ago. You're just following in a long tradition of human preparedness. It is clearly the sensible thing to do.

About the Authors

Cam and Michelle Mather live independently off-the-electricity-grid near the village of Tamworth, north of Kingston Ontario. They use the sun and wind to power their home and business, Aztext Press. They publish books and produce DVDs about renewable energy and sustainability. They focus their energies these days on running a CSA or Community Supported Agriculture which supplies 40 families with a weekly box of organic fruits and vegetables from their gardens.

They have been featured in a number of newspaper, magazine and television profiles and the City of Burlington's Sustainable Development Committee produced an educational video about the Mather family called "Changes: a low impact environmental lifestyle." The video won a number of awards.

Cam and Michelle contribute regularly to Mother Earth News and other magazines.

They have produced DVDs on organic vegetable gardening and installing a home-scale wind turbine. Cam's first book, Thriving During Challenging Times, the Energy, Food and Financial Independence Handbook, was published in 2009. He put his 35 years of organic and market gardening experience to use in his next book, The All You Can Eat Gardening Handbook: Easy, Organic Vegetables and More Money in Your Pocket. Cam and Michelle have also shared their experiences in moving off the electricity grid in their 2011 book, Little House Off The Grid, Our Family's Journey to Self-Sufficiency.

They are striving to make their home "zero-carbon" and with their extensive garden they aim to be completely food and energy self-sufficient. Through Cam's workshops at colleges and universities he has motivated thousands of participants to invest in energy efficiency and renewable energy. His blog about country life is enjoyed by thousands of readers and is republished on the Mother Earth News website.

You can learn more about Cam and Michelle and follow their blog at **www.cammather.com**.

Also by Michelle and Cam Mather

One family's journey from a comfortable life in the city to living sustainably off the grid in a century-old farmhouse.

Little House Off The Grid

Our Family's Journey to Self-Sufficiency

Michelle & Cam Mather

220 pages 6" x 9"

ISBN 978-0-9810132-5-1

$19.95 Cdn/US

Like so many other city-dwellers, Cam and Michelle Mather longed for a simpler, quieter life in the country. When they found a century-old farmhouse on 150 acres of land that was in their price range, they jumped at the chance to make their move. The fact that the home was "off-grid" with no power or telephone lines connecting it to the outside world seemed like a bonus!

Twelve years later their life in the country is not quite as simple as they had envisioned but it is peaceful. There were more challenges than they could have anticipated as well as more rewards.

Along the way they installed more solar panels, erected a wind turbine and upgraded and replaced all of the major components of their off-grid electrical system. They installed a solar-thermal hot water system, figured out how to have a phone, internet and satellite TV and kept their home heated with wood cut from their own property. They also carved out a garden and began growing much of their own food.

They acquired many new skills and knowledge but most importantly they learned to appreciate the value of good neighbors, good books and good manure.

Read our blog, keep informed about upcoming workshops and follow what a fellow 'sensible prepper' is up to on a daily basis at www.cammather.com.

Also by Cam Mather

Grow your own organic vegetables and enjoy a "One Hundred-Foot Diet!

The All You Can Eat Gardening Handbook

Easy Organic Vegetables and More Money in Your Pocket

Cam Mather

260 pages 8" x 10"

ISBN 978-0-9810132-2-0

$24.95 Cdn/US

While many books make vegetable gardening look difficult with charts and checklists and talk of trace minerals and hard to find soil supplements, growing vegetables can seem intimidating. That's why ***The All You Can Eat Gardening Handbook*** is such a breath of fresh air. It assures readers that there's nothing to it, and encourages them to just get out there and do it. With basic tips and techniques it provides enough tools to inspire gardeners but doesn't overwhelm them.

The North American diet uses lots of fossil fuels and as we run out of the easy oil we will spend an ever-increasing percentage of our incomes on food. With many hard hit by the economic crisis, growing your own food simply makes economic sense. ***The All You Can Eat Gardening Handbook*** examines the health benefits of each of the vegetables and fruits listed. Sure, Grandma always told us to eat our vegetables but as adults it's nice to know about all the incredible health benefits of each item you're growing. The book also provides strategies for harvesting rainwater, watering with drip irrigation and dealing with some of the challenges our changing climate may throw at you.

Whether you have a small lot in the city, a suburban backyard or a large country property, ***The All You Can Eat Gardening Handbook*** is the tool you need to get motivated to start growing healthy, local, inexpensive and organic vegetables and eating the 100-Foot Diet today!

DVDs to help you in your quest for independence

Grow Your Own Vegetables

With rising fuel and food costs, this 2 hour DVD provides everything you need to turn your backyard into your own personal produce department. This program covers soil preparation, starting seeds, planting, weeding and watering, dealing with pests, and harvesting and storage of your bounty

ISBN 978-0-9733233-9-9

Home-Scale Wind Turbine Installation

This video is a step-by-step guide to putting up a home sized wind turbine using a common tubular steel tilt-up tower and winch. From evaluating your location, installing anchors, wiring, assembling the tower and using the winch to properly raise the tower, this DVD will guide and inspire your move to green energy.

ISBN 978-0-9810132-0-6

Living with Renewable Energy

This 2 hour long DVD is a tour of several off-grid homes, including the home of Cam & Michelle Mather. It shows how to enjoy a typical North American lifestyle, powered by the sun and wind.

ISBN 978-0-9733233-8-2

Read our blog, keep informed about upcoming work-shops and follow what a fellow 'sensible prepper' is up to on a daily basis at www.cammather.com.

Made in the USA
Charleston, SC
25 July 2015